J. H. LAMBERT

COSMOLOGICAL LETTERS
ON THE ARRANGEMENT OF THE WORLD-EDIFICE

Translated with an Introduction and Notes by

STANLEY L. JAKI

Science History Publications
New York

First published in the United States by
Science History Publications
a division of
Neale Watson Academic Publications, Inc.
156 Fifth Avenue, New York 10010

First Edition 1976
Designed and manufactured in the U.S.A.

Library of Congress Cataloging in Publication Data

Lambert, Johann Heinrich, 1728-1777.
    Cosmological letters on the arrangement of the world
edifice.

    Translation of Cosmologische Briefe über die Einrich
tung des Weltbaues.
    Bibliography: p.
    Includes index.
    1. Astronomy--Early works to 1800. I. Title
QB42.L25   523.1   75-31623
ISBN 0-88202-042-0

ACKNOWLEDGMENTS

It is my pleasure to record here as a token of my appreciation the names of the following: Martin Steinmann and Rolf Stoeckling, archivists at the Universitätbibliothek Basel, for their obliging assistance in my study of Lambert's correspondence; Bernhard Ulmer, Professor of German Literature in Princeton University, for his generous advice concerning the peculiarities of Lambert's style; Rosemary H. Smith and Anne S. Williams for their untiring help with proofreading; Esther Breisacher for typing the manuscript; B. M. Israel of Amsterdam for obtaining rare printed material; Mary Smyth, Librarian of the Royal Scottish Observatory of Edinburgh, for providing photographs of the original printed here; and last but not least Neale W. Watson, President of Science History Publications, for his keen interest in the project.

The name of Seton Hall University registers my appreciation to too many there to mention each one individually.

S. L. J.

Contents

INTRODUCTION

I Lambert's Life and Work

On an evening in March 1764, a prospective member of the Berlin Academy of Sciences was introduced to Frederick the Great, its royal patron. The scene could not have been more unusual. The candidate was seated and almost all the candles were extinguished shortly before the king entered. "Would you do me the favor of telling me in what sciences you are specialized?" the king asked the visitor of whom he could at best see a dark silhouette. "In all of them," came the answer from the man in the dark. "Are you also a skilful mathematician?" the king asked again. "Yes," the visitor answered. "Which professor taught you mathematics?" the king pressed on. "I myself," went the reply as curt as before. "Are you therefore another Pascal?" "Yes, Your Majesty." At that the king turned aside, for he could hardly hold his laughter, and returned to his private office. Later, at dinner, the king remarked that the greatest blockhead in the world had just been presented to him for membership in the Academy.

About a year later, the blockhead, Johann Heinrich Lambert, became a member of the then most prestigious scientific body in the world. There came after all a happy ending to what may first appear a mere comedy, or perhaps an anecdote with little or no foundation in fact. Yet, such was the first encounter between Frederick the Great and Lambert, one of the great among self-taught scientists.[1] Fourteen years later the incident was respectfully recalled in a somewhat shorter form by Jean L. S. Formey, perpetual secretary of the Academy, in a eulogy[2] which gives the impression that soaring rhetoric for once was not a hollow sequence of words.

In the four curt replies which Lambert gave the king in that darkened stateroom of the Royal Palace there is condensed the whole drama of Lambert's life. The room had to be darkened at the insistance of Lambert's champions at the court. They feared that the king would be repelled by Lambert, whose strange appearance made some think that he was a visitor from the moon.[3] The scarlet red tailcoat, the turquoise blue vest, the black trousers and white stockings, which he was wont to wear, were odd enough.[4] Even more so was the similarity to a block of a face dominated by a large, straight forehead and by a protruding broad jaw. Moreover, this face was always held sideways whenever Lambert spoke to anyone. This was one of the strange mannerisms of a man who was his own teacher not only in science but also in etiquette.

1

His curt replies to the king were not meant to be a discourtesy. In part they evidenced his refusal to conform to the manners of refined society. He wanted to remain a plain man, as primitively straightforward in his speech as in his preference for simple food. Although a member of the Academy, he loved to visit the ordinary taverns of the capital, to enjoy plain meals, mix with simple folk and give free rein to his bellowing laughter.[5] His curt replies also had to do with his esteem for simplicity which he thought to be the duty of a Christian. In the Berlin of Voltaire he created great consternation by taking part in the communion services of the Reformed Church which he regularly attended.[6] The curtness of his answers also evidenced the self-assurance of a self-taught genius, an assurance that was only strengthened by Lambert's isolation from university life. In that self-assurance nothing could shake him, not even his disastrous first visit with Frederick the Great. Shortly afterwards, when a friend tried to cheer him up and counseled patience in view of the king's heavy schedule, Lambert's reply was plain self-assurance: "I have not the least uneasiness on the subject; his [the king's] glory is at stake, and should he not name me, it would be a blot in his own history."[7]

Lambert proved right. The king appointed him, referring to his learnedness which seemed to have no bounds.[8] His unique privilege to read papers in all divisions of the Academy attested his encyclopedic wealth of information. He acquired it with no more formal help than what could be had in the first six grades of school in Mulhouse. He was not yet twelve when he had to lend full help to his father, a struggling tailor, whose family came to Mulhouse with other Huguenots more than a century before Johann Heinrich was baptized on August 29, 1728. In addition to an older brother, he had two sisters and three younger brothers. Mulhouse and its surroundings were at that time associated with the Swiss Confederation, and Lambert could naturally be taken for a Mulhousino-Helvetus.[9] In Lambert it was their fellow countryman whom Euler and Sulzer, Swiss members of the Berlin Academy, tried to add to its roster. Mulhouse had one school and several parishes with a number of graduates of theological academies serving as pastors. Forty years later, Formey remarked, not without some touch of malice, that they were theologians who thought that theology was the only study worth pursuing.[10] One of them tried to obtain a scholarship in a theological academy for the boy whose learning obviously transcended what was offered in the local school.

Some of the particulars that survived about young Lambert's self-tutoring are not without poignancy. They show a boy selling his small pencil drawings to other children for a penny or two to let him buy books and candles which his family could not afford. The family home might have indeed perished in flames had the young boy, reading late into night, not discovered that embers left in the fireplace were setting the kitchen afire. What the young boy studied most in those late hours was his most cherished purchase, a booklet on practical mathematics and geometry. He learned it so well that he was able to discuss matters with carpenters coming to repair the house, one of whom was so impressed that he offered him his own book on arithmetic and geometry.[11]

Young Johann got his spurs. A year or so later he was busy calculating the orbit of the famous comet of 1744, which in the French part of the Swiss Confederation aroused the interest of another precocious youngster, Jean-Philippe Loys de Chéseaux, best remembered today for his essay on what later became known as Olbers' Paradox.[12] Fourteen years later Lambert gave the only discussion of that essay during the eighteenth century in his *Photometria,* which became a classic of science.[13] The book was one of the fruits of his twelve-year stay in Switzerland, where his first station was Basel. He went there in early 1746 at the recommendation of J. H. Reber, town clerk in Mulhouse, to serve as secretary to J. R. Iselin, editor of the *Basler Zeitung,* who also gave various courses at the University and later became professor of law there. Instead of attending Iselin's classes, which he was free to do, and of availing himself of some formal training, Lambert preferred to teach himself.

A long letter of his, written on December 6, 1750 to Pastor Riszler in Mulhouse,[14] reveals that his fondness for mathematics and physics led him to a deeper study of the laws of human reasoning and of the criteria of valid knowledge. Another motivation was his Christian concern with perfection and happiness. After recalling that during the four years prior to his coming to Iselin he had mastered Latin and French, he noted that his work as secretary took up only half of his time.

"I, therefore, got hold of a few books to learn from them the first principles of philosophy. I have soon found that the primary project of my efforts should be the finding of the method of making myself perfect and happy. I have, however, also realized that the will perverted by nature cannot be improved, if reason is not first freed from biases and is not properly enlightened. This was also my first observation, and I have found, in what Wolff pointed out [in his book] on the power of human understanding, Malebranche [in his book] on the investigation of truth, and Locke in his thoughts on human understanding, those rules which were for me of great profit in learning about reason itself and about its shortcomings, as well as in searching for the truth. This profit first evidenced itself in connection with [my study of] the mathematical sciences, and especially algebra and mechanics, which gave in my hand clear and basic examples by which I could reinforce the rules learned so far, and let them become, so to speak, my flesh and blood."[15]

As Lambert noted in the same letter, he had found no reason yet to regret this program of self-instruction. It not only helped him to instruct himself more effectively in other sciences but also his religious efforts benefited.

"I know only too well that the will, in general, desires the good and shirks the evil; but I also saw that all this presupposes that one must first know clearly what is truly good and what is truly evil, so that we may not take apparent goods for true ones and that we may not let ourselves be blinded by Satan, by the world, and by our own passions. Therefore I did not abandon studying the moral teachings of the Holy Scriptures, and as I came afterwards to reading Puffendorff's booklet on the duty of man and citizen, besides other philosophical ethics, I had the opportunity to perceive quite clearly the advantage of divine [Biblical] ethics over the others and to devote myself to it all the more earnestly."[16]

Not exclusively though. In the next breath Lambert noted that since the principles of philosophical ethics show the road to what is right by nature, he was not "to lose sight of a better knowledge of them. In particular, I have availed myself in this respect of the counsel which Rollin gives in his *Méthode d'étudier et d'enseigner les belles lettres* to those who want to devote themselves to the liberal arts."[17]

When Lambert wrote this letter he had already been for two years in Chur, tutoring in the house of Count von Salis. His charges were the thirteen-year-old Anton von Salis, a grandson of the count, Baptista, a cousin of Anton, also thirteen, and the nine-year-old Johann Ulrich, a distant cousin. His success in instructing them served Lambert as a justification of his lack of regular training. "It is true," the same letter continues,

"that in reading this and other books I have sufficiently felt the lack of oral instruction and I had to leave unanswered some questions that occurred to me in connection with those subjects; but I considered that I have supplemented that lack with all the greater diligence and I have, with God's help, progressed so far that I was able to achieve over the material which I had learned a completely satisfactory mastery, so that I hope to make a journey within three years with two young gentlemen, whom I instruct in the languages, in Osterwalder Catechism, in arithmetic, geometry, military engineering, geography, and history, to the University of Utrecht and from there to England and France."[18]

No wonder that the letter of Lambert, he was now twenty-two, came to a close with his expression of gratitude to the Almighty about his good fortune and prospects.

Happily for Lambert, he started out on his travels not in 1753 but in 1756. His three more years as tutor in Chur[19] gave him further time to train himself in philosophy as well as in mathematics, physics, and astronomy, and to start with his own original investigations, both speculative and experimental. His progress can be followed closely from his Monatsbuch, a sort of condensed diary, in which from 1752 on he listed in a few lines at the end of each month the work done.[20] The material of several of his major books was collected during those years. Since Count von Salis was a prominent figure, many important visitors came to his residence, and Lambert, treated by the Count and his family with invariable courtesy, was able to make himself known to Swiss scientists, several of whom quickly recognized his genius. It was at their invitation that he published in 1755 in the *Acta Helvetica* his first paper, a seventy-page-long essay on the expansive force of heat in which the topic was treated in a rigorously mathematical manner in eighteen problems amplified with solutions and demonstrations.[21] The essay showed Lambert in full command of geometry and calculus, and produced a very favorable impression. One-fourth of the next or third volume of the *Acta* was taken up by three articles of Lambert which in turn showed him as a pure mathematician, a theoretical physicist, and a careful collector of observations.[22] Moreover, the Preface of the volume in question was a long and flattering invitation to Lambert to study further the possible influence of the moon and sun on the atmosphere.

The "Hofmeister," as Lambert liked to refer to himself, was already a member of the Literary Society of Chur and of the Swiss Scientific Society of Basel, when he started out in the spring of 1756 on an almost three-year-long trip with Anton and Baptista, now eighteen and ready to have a taste of university education. Lambert was even more eager than his charges to be near a good university. The first stopover on the journey was Göttingen, perhaps the finest university then anywhere. Lambert became friends with two prominent professors there, A. G. Kaestner, a physicist, and Tobias Mayer, an astronomer. He left Göttingen in late summer as a corresponding member of its Learned Society. The rest of 1756, the whole year of 1757, and the better part of 1758 were spent in Utrecht, with visits to various Dutch cities. Pieter van Musschenbroek, when Lambert first visited him in Leyden, thought that he was to instruct an itinerant student, but before long he found a teacher in the stranger. Shortly afterwards Lambert published in The Hague in 1758 his book on the path of light in air and in various media.[23] The small work was a tightly knit sequence of propositions, proofs, theorems, lemmas, and data of observations. If it is true that he had discovered the exponential decrease of the intensity of light in various media independently of Bouguer, then Lambert's reference to Bouguer in the preface[24] might have perhaps been due to the suggestion of his publisher or of some friend. In the summer of 1758 Lambert set out for Paris where he met d'Alembert and found in Messier a kindred spirit. By October he was back in Chur with considerable gain in experience and fame.

Lambert's next work, published in 1759 on perspective,[25] was a goldmine of intuitive insights valuable to artists. A year later his *Photometria* served massive evidence of his originality.[26] This streak of originality appeared undiminished in the two books which he published in 1761. One was a discussion of the properties of cometary orbits, containing a new method of calculating the path of comets.[27] The other was his *Cosmologische Briefe* or *Cosmological Letters.* On a cursory look it seems to have been motivated by his preoccupation with comets, but its true wellspring was a vision of the universe as a totality of hierarchically ordered systems of stars. Order and hierarchy were synonymous for Lambert with purpose, and this in turn was one of the basic truths that in Lambert's eyes were self-evident and provided a necessary framework for valid reasoning. Apart from his interest in obtaining membership in the Berlin Academy, Lambert must have composed with no small relish his essay on the method of proving more correctly the propositions of metaphysics, theology, and ethics. He hoped to win with it the Prize set for 1761 by the Academy.[28] Another participant in the contest was Kant, who saw a most congenial spirit in Lambert after the publication of Lambert's first major work on the method of securing sound knowledge.[29] It was the *Neues Organon,* "or thoughts on the exploration and identification of truth and on its distinction from error and appearance," published in 1764.[30] The 9th of January, 1765, witnessed Lambert's appointment as member of the Academy with the privilege of presenting papers in all its classes. Two weeks later he gave his Antrittsrede, "Sur la liaison des connaissances qui sont l'objet de chacune des quatre classes," a fitting evidence of the universality of his interests.[31]

During his twelve years as member of the Berlin Academy Lambert produced about 150 scientific papers. A friend of his rightly called him a "true dissertation-machine."[32] In fact, even during his daily walks Lambert often spoke as if he were composing a paper. Many of those papers showed his ability to tackle concrete problems and perform complicated calculations with ease. As a recognition of his contributions Lambert in 1771 was appointed Oberbaurat, or chief counsellor of constructions. But today he is not remembered for his solutions of problems in civil engineering, for his suggestions to improve precision instruments, for his efforts to organize international co-operation among astronomers and observatories, or for his skill to evaluate the explosive force of gunpowder. His enduring fame as a scientist does not even rest on his pioneering studies in the physics of light, the *Photometria,* and the *Pyrometrie,* a collection of his studies on heat published two years after he died on September 25, 1777. His really lasting achievements were in pure mathematics and geometry, among them his proof of the irrationality of e, the basis of natural logarithm, and of π, the ratio of the circumference of a circle to its diameter.[33] He came within striking distance of formulating a non-Euclidean geometry. He himself saw his own forte in mathematics. To a question about the greatest living mathematicians Lambert replied: "In the first rank are Euler and d'Alembert, the second is Lagrange; I say, now, because he will soon catch up with the first two. I am the third. Further I do not go, because I know of no one who could still be added."[34]

His creativity in mathematics witnessed a mind given to penetrating vision which evidenced itself even in fields that could appear far removed from his ken. Historians would find ample food for thought in his uncanny evaluation in 1770 of the political situation of Europe, of its turbulent future, and of the impending rise of two superpowers, America and Russia.[35] The same visionary power was also the real force behind his view of the cosmos. Today, when there is so much interest in analyzing the working of scientific minds, Lambert's reasoning in the *Cosmologische Briefe* could form a most interesting subject of study.

From the viewpoint of the history of science, there is more than one strong reason for a careful analysis of this work. The *Cosmologische Briefe* is possibly the least read and the least accessible scientific classic.[36] No wonder that very generic or mostly erroneous notions about Lambert's cosmology are standard fare in secondary literature. Apart from helping to reduce the vast number of incorrect notions present in most histories of science, the study of the *Cosmologische Briefe* provides a fascinating possibility for seeing the history of cosmology at one of its most memorable transitions. Lambert's speculations constitute illuminating material not only when set against the cosmological dicta of his older contemporaries, but especially when seen in the light of what was achieved by Herschel, ten years his junior. Finally, there is the present-day interest in hierarchical cosmologies, whose champions are all too eager to speak of Lambert, but find it difficult to form a sound idea of what he had actually written on the subject.

II   The *Cosmologische Briefe*

A)   ITS ORIGIN

It appears that the *Cosmologische Briefe* owed its origin to a sudden impulse. An entry in June 1760 in Lambert's Monatsbuch, "I have started a series of letters on the system of the world," is the first written indication of his involvement with cosmology.[37] Five years later he revealed in a letter to Kant that his first and sole sketch used in writing the *Cosmologische Briefe* dated back to a summer night in 1749. The sketch written in a single quarto sheet contained the explanation of the visual appearance of the Milky Way as a realm of stars confined to a space resembling a flat disk.[38] Reference to that single quarto sheet again appeared in Lambert's letter of March 7, 1773 to J. L. Boeckmann.[39] To him Lambert also confided that in writing the *Cosmologische Briefe* he did not have his books with him and because of this he forgot to use several data that could have strengthened his argument. They were the Magellanic Clouds, the explanation by Maupertuis of the various visual shapes of "nebulous stars,"[40] and some articles in the *Mémoires* of the Academy of Sciences in Paris on the pale light in Orion.[41] But as Lambert makes it very clear in Letter XII, the full development of his ideas on the Milky Way and the systems of stars was a very gradual, almost unnoticeable process.

In June 1760 Lambert was already in the second year of a five-year phase of his life that saw him move restlessly in search of a suitable academic position. In March 1759 he took leave of the Salis family and most likely left with them much of his belongings, including his books, for safekeeping. In April he was in Zurich where he tried, without success, to spot Halley's comet with a telescope, the return of which had been greeted with much excitement.[42] From June to September he visited with his mother and other members of his family in Mulhouse. From September on he was in Augsburg where he lodged in the house of G. F. Brander, the famous maker of precision instruments, with whom he kept a life-long correspondence.

His moving to Augsburg was motivated by his impending appointment to the fledgling Academy in Munich, a city which was too Roman Catholic for Lambert to take up residence there.[43] Augsburg, where Protestants and Catholics were in balance, provided Lambert with a more congenial atmosphere. There, he felt, he could live entirely in his own thoughts which at that time were dominated by comets. The *Cosmologische Briefe* was published almost simultaneously with his book on the "more outstanding properties of the orbit of comets."[44] Undoubtedly the return of Halley's comet played a considerable part in prompting him to compose a cosmology in which comets were most prominent. Another reason might have been the advisability of publishing as much as possible and in the shortest time in order to obtain membership in an Academy more prestigious than the one in Munich. Yet, as will be seen shortly, Lambert, in his efforts to secure a post in the Berlin Academy, did not mention his *Cosmologische Briefe*. He started work on it in June 1760, but his attention was also drawn to several other projects.[45] As his letter to Boeckmann also attests, the *Cosmologische Briefe* was composed piecemeal, one Letter serving as the occasion for the next.[46] As a result the work is not without redundancy. This can also be due to the leisurely exchange of ideas which Lambert wanted to have between two correspondents—himself and an imaginary friend of like mind. The Letters of uneven number were supposed to be those sent to him, whereas the even-numbered Letters were his replies. Yet the difference is anything but striking at times.

As the Monatsbuch states, he continued work on the *Cosmologische Briefe* in July.[47] The lack of reference to it in August may be an indication that by the end of July Lambert might have completed Letter XIV where, as he noted in the Preface, he originally wanted the whole work to close.[48] The next entry in the Monatsbuch about the *Cosmologische Briefe,* again referring to continued work on it, dates from September, or a year after his arrival in Augsburg. Lambert's other reason for moving to Augsburg was the publisher, the widow of Eberhard Klett, who was just bringing out his *Photometria.* By October the *Cosmologische Briefe* was completed[49] and in January 1761 the manuscript was handed over to Klett's widow who had it printed in two months' time.[50] The haste is reflected in the large number of typographical errors of which the garbled sequence of pages in the latter part of the Preface is the most conspicuous.[51]

B) ITS CONTENTS

The lengthy PREFACE which introduces the *Cosmologische Briefe* certainly reveals Lambert's full awareness of its place in his program of publications.[52] From the astronomical viewpoint the work is, according to him, a development of what he had briefly said on stars in his *Photometria.* From the philosophical viewpoint it is a preparation for his systematic discussion of epistemology, and especially of the role of teleological considerations. The conspicuous aspect of teleology is the postulate that every part of cosmic space should be suitable for maintaining life. It is this consideration that justifies the filling of all usable space around each sun with comets which should be considered as bodies differing from planets only in the structure of their atmosphere and in the shape of their orbit. Another important feature of a purposefully arranged cosmos is the stability of their major constituent parts, such as planets, comets, stars, and systems of stars. Since all celestial bodies have the purpose of supporting their inhabitants, they cannot undergo major transformations either with respect to their orbits or to their physical constitution. The "evolutionary" view of the universe is, Lambert warns his reader, contrary to teleological principles, as is also, partly for the same reason, the notion of the infinity of the universe.

Whereas Halley's Table of comets gives considerable observational evidence to the "teleological" filling with comets of the spaces of the solar system, the observational evidence is rather meager when, in the second half of the work, attention is turned to the teleological arrangement of that part of the cosmos which lies beyond the solar system. The teleological arrangement in question is a partition of stars into ever more inclusive systems resembling the ordering of planets and comets around the sun. Evidence in support of this view is not completely absent. On the basis of observations it should seem most likely that the realm of stars visible to us is confined within a shape resembling a flat disk. Also some recent conclusions about the displacement of some fixed stars should make it appear very plausible that stars are also in motion. Such a displacement should be further corroborated by direct observations, and this invites some remarks about the mutual connection between a priori postulates and a posteriori verification. There follows the admission on the part of the author that the validity of his arguments will therefore be largely dependent on the mental viewpoint of the individual reader. The reader is warned that experimental evidence is far from being complete about the partition of the Milky Way into smaller systems and about the orbiting of stars in each such system around a central body. Still, the reader is cautioned against taking a sceptical view. Cosmology is a science not only of the parts but also of the whole. Historically, Copernicus made only the first step in going beyond Ptolemy, but while these Letters intend to show the direction which post-Copernican and post-Newtonian cosmology should take, it is recognized by the author that his style is no match to that of Fontenelle, who discoursed with so much liveliness about the inhabitants of other planets. Finally, justification is given for the letter-form together with some hints about the respective character and position of the two friends engaging in this exchange of letters.

The FIRST LETTER starts with a reference to the return of Halley's comet in 1759, and almost immediately the question is raised whether comets are harbingers of disaster. That the cosmos is full of cosmic catastrophies due to collisions of celestial bodies has been the favorite theme of philosophers, of whom Burnet and Whiston are mentioned. Such a catastrophic view of the cosmos is opposed in the SECOND LETTER with a view to the many evidences which show that on the earth living beings of all forms have a habitat suited for their conservation. This teleological view is firmly upheld also in the case of larger habitats such as celestial bodies, and the advocates of collisions are taken to task for their inability to predict them. It is argued that collisions can only occur if the paths of celestial bodies are carefully adjusted to that end and that considerations of probability heavily oppose the assumption that the satellites of Saturn and Jupiter are comets captured by these great planets. Consequently, while small displacements in the position of planets and comets are admissible, any major dislocation of a celestial body from its orbit should be considered most unlikely.

After a brief review of the foregoing considerations, the THIRD LETTER strengthens the non-collisional view of the cosmos by a presentation of the unlikelihood that a comet could have ever become our moon. Still it is noted that the application of teleology is more difficult with respect to entire celestial bodies than in reference to details of our own habitat. It is rather difficult to infer purposefulness in a specific form from starlight and from gravity, universal as they are. The case is quite different, however, when Halley's Table of comets is analyzed. Since the twenty-one comets listed there have not been arbitrarily selected, their orbital features can be used for a "statistical" evaluation which gives hints about the manner in which the system of celestial bodies, of which comets are the most numerous, should be teleologically built up. For one, comets that are placed close to the sun should have a steep angle of inclination. For another, most comets should be put either into elliptical orbits with long periods, or into hyperbolic orbits that exclude the possibility of return.

The building up of the world of comets continues in the FOURTH LETTER but only after the admission that the notion of a collision-free world is ultimately anchored in considerations about the wisdom of the Creator. It is now claimed on the basis of Halley's Table that the number of comets increases with the square of distance from the sun. Various reasons are given why of the 3600 comets estimated to have their perihelia within Saturn's orbit only 90 at most are visible. It is then inferred that the number of comets with perihelia within the orbit of Mercury might be five million. It is also part of the perfection of the universe that elliptical cometary orbits can readily turn into parabolas and hyperbolas. Comets whose orbits undergo the latter change can be pictured as being on a journey from star to star. Their inhabitants have therefore the finest opportunity to collect astronomical observations about the whole universe, a possibility which implies that they are structured both in mind and body for vast units of space and time.

It is acknowledged in the opening part of the FIFTH LETTER that the complete verification of such details demands enormous spans of time, but doubts are once more allayed by reference to a "total" view steeped in cosmological (teleological) considerations. The difficulty is then aired about the physical constitution of the denizens of comets, who must be able to withstand freezing cold in cosmic places very remote from any star, and sizzling heat in their vicinity. Does this not presuppose that comets are very different from planets? But the chief difficulty seems to be with the accommodation of so many comets around the sun without exposing them to collisions. A model based on the spokes of a wheel is proposed but not developed at this point. Rather the fullness of every bit of matter with life is emphasized, and warnings are sounded against thinking about life in a too anthropomorphical manner. It is on such ground that the image of super-astronomers traveling on comets from sun to sun is again evoked.

The notion of the fullness of life across the universe and the "general considerations" as its basis form the starting note of the SIXTH LETTER. It is argued that in the universe space and time increase together, which serves as another warning against thinking too narrowly about possible forms of life. Specific considerations are given to conditions of life on the comet of 1680, about which it is suggested that the maximum heat to which it was exposed at its perihelion had to be 2000 times greater than that of glowing iron. Its inhabitants were protected by its atmosphere which swelled up only in the vicinity of the sun. This is why comets are not visible at great distances, a point which introduces further reasons to explain why only a few comets are visible. These reasons are based on the analysis of cometary orbits. Fairly rounded orbits remain always at great distances from the sun and from the central regions of the solar system. It is also shown with a model resembling an armillary sphere that such cometary orbits cannot occur in great numbers if collisions are to be avoided. Quite different is the case with very elongated orbits. Now the spokes-in-the-wheel model is developed in some detail and explanation is given of the fact that comets far outnumber planets.

The very great number of celestial bodies, especially of comets, is the opening topic of the SEVENTH LETTER, in which attention is given to the correlation of the system of comets with planetary orbits. Once more the chief concern is the avoidance of collisions. The specific problem is the manner in which planetary and cometary orbits can intersect one another. Because of the paucity of cometary orbits already calculated, emphasis is given to teleological reasons and this again brings up the presence of life everywhere in the cosmos and in most varied forms. The various degrees of light and heat are discussed, and further speculations are offered about the atmosphere of comets as protective umbrellas for their inhabitants. It is in this connection that an explanation is attempted about the development of the tail of comets. The explanation which implies some loss of material for the comets is then balanced by efforts to reduce that loss to a minimum and to provide for the replenishment of the lost material. In a teleological perspective comets must be enduring, stable entities.

In the EIGHTH LETTER it is once more emphasized that the truth of all this largely depends on one's acceptance of the teleological view which implies the existence of living and thinking beings everywhere in the universe. It is in this light that the various steps of filling cosmic spaces with comets are reviewed and the Newtonian law of gravitation is extolled as being in full harmony with cosmic teleology. Reference to that law leads to a review of what had already been said about Halley's Table with further reflections on the regular arrangement of planets. This arrangement, so contrary to being the effect of chance, provides further support to the teleological view. Particular emphasis is given to the even distribution of the sun's heat through its rotation and to the maintaining of water on the surface of a comet with the aid of its protective atmosphere. This maintaining of steady conditions is necessary because germs of life would not survive a catastrophic transformation of a comet into a planet. Yet, it is recognized that one can only be sure about the very general features of the cosmos.

A complete knowledge of all details presupposes, as noted in the NINTH LETTER, the completion of Halley's Table. On its basis it is now suggested that 12,000 comets may very well have their perihelia inside the orbit of Saturn. Within a sphere ten times larger there could therefore be well over a million comets. Then attention is turned to the system of planets, and estimate is given of the extremely small probability that all its sixteen members should move, because of chance, in the same direction. Following some remarks about the connection between Newton's law of gravity and the orderly motion of the solar system, the question is raised whether the same order can also be found among the fixed stars. Immediately the Milky Way is mentioned with the question whether its visual appearance is due to a greater proximity of stars to one another or to many rows of stars behind one another. Finally the question is asked whether the laws of motion of the system of fixed stars shall ever be found.

In the opening part of the TENTH LETTER the laws of motion support the proposition that without orbital motion all systems would collapse into their center of attraction. Attention is now turned to a comparison of ancient and modern star catalogues as a possible source for verifying the displacement of at least some stars. The next question is whether the motion of stars is around a common geometrical center of gravity or around a very massive central body. Soon, however, the subject matter is again the explanation of the visual appearance of the Milky Way as a realm of stars confined within a space "not spherical but flat, roughly like a disk, whose diameter is many times larger than its thickness." The Milky Way is then said to be composed of many smaller systems of stars, all lying in the same plane. The visible stars, including our sun, belong to a system different from those constituting the Milky Way. The question is left open whether the Milky Way belongs to still greater systems. Once the enormous number of stars in these systems is emphasized, it is argued that the sun must be fairly close to the center of the star system to which it belongs.

The broader and deeper sense of "Copernicanism" is brought up in the ELEVENTH LETTER. It consists in the subordination of all stars into more and more inclusive systems, all of which move around a central point or body. This in turn raises the question whether such subordination can be detected in the Milky Way. To decide this point it has first to be ascertained whether stars in the Milky Way are at least as distant from one another as are the nearest stars from us. The close packing of stars in the Milky Way is rejected as contrary to the notion of suitable habitats. Then a model, consisting of rows of lamps, is outlined for the explanation of the visual appearance of the Milky Way, and it is noted that this is the first correct explanation ever, including ancient times. The model implies a considerable spacing between the stars, and complies with the principle that the light of any star is to be utilized by many planets and comets around it. Such a teleological consideration must decrease the weight of some irregularities, such as the uneven contours of the Milky Way. The truth of teleological cosmology rests on the interconnectedness of its various aspects and on the prediction on its basis of various phenomena.

The TWELFTH LETTER continues on a note which brings Lambert himself into the foreground as he recalls the moment when he first perceived that the Milky Way was a realm of stars confined within a disk. He now also discloses that it was in the same context that he was seized with the idea that the Milky Way itself was made up of systems of fixed stars. Then Lambert states that the ultimate truth of his cosmology depends on a reliable estimate of the mutual distance of stars. Various possibilities of coping with that task are reviewed and found wanting. The necessary and orderly motion of celestial bodies is emphasized as the chief reason why the stars in the Milky Way should be paced at sufficiently great distances from one another. Then it is stated that due to geometrical reasons our system of stars can only be surrounded by six systems of stars similar in size, lying in the same plane. It is also asked whether the pale light in Orion can be taken as the nearest Milky Way. The final remarks are about the actual shape of the Milky Way and about the force of gravity which should emanate from the massive center of our system of stars and which should be felt by each member of our solar system. Are some anomalies in the motion of planets due to that force?

In the THIRTEENTH LETTER mention is made of the collapse of the Ptolemaic system and with it of the sphere of fixed stars. Consequently, the motion of stars in free space and their unequal distances from us become inescapable assumptions. It is suggested that observation of some stars might be particularly useful in gathering direct evidence of stellar motion. Even more hope is attached in this respect to careful investigation of various anomalies in the motion of the solar system. Is, for instance, the motion of the nodal points of the planets along the zodiac a consequence of the sun's motion toward other stars? An estimate is made of the velocity of the nearest stars around their center on the assumption that they had changed their position by $\frac{1}{4}$ degree since Hipparchus. The question is then raised about distinguishing stars in the Milky Way by telescopes, about the distribution of stars according to brightness and about the position of our system of stars with respect to the Milky Way. .

The FOURTEENTH LETTER starts with answering the question of resolving stars in the Milky Way with telescopes. In particular it is explained why the telescopes do not show the stars as perfectly round figures. The explanation depends both on the way in which telescopes function and on the structure of the eye. A consequence of this is that one has therefore to assume many stars to be thousands of times farther away than stars of the first magnitude. The estimated number of stars in our own system leads to a brief estimate of the "inner diameter" of the Milky Way. That its outer diameter cannot be ascertained by telescopes is ascribed also to the weakening of starlight in the ether. Then several circumstances are specified about anomalous phenomena in the motion of planets that might be due to their gravitating toward the center of the local system of stars. Possible periodic changes in the earth's orbital velocity are singled out as the area where further research might reveal evidences about the motion of the center of our system in the direction of Orion.

In the opening of the FIFTEENTH LETTER gratitude is expressed for the explanation of the appearance of stars through the telescope and for the suggestion concerning the anomalies in the dynamics of the solar system. Then it is reported that Professor Mayer of Göttingen had found evidence for the displacement of stars. Since this proves the motion of stars, the hitherto synthetic procedure can be replaced by an analytical one. Analysis of possible forms of motion shows that an orbit produced by a central force is the only one compatible with an orderly and purposeful arrangement of the cosmos. Newton's law of gravity is then extended throughout the universe, and it is suggested that the pressure of the ether might be the cause of gravity. But it is deemed to be more profitable to return to the specifics of the solar system and especially to Halley's Table. Its data are analyzed in terms of the six parameters that determine the orbit of comets. The general conclusion is that their number increases with the square of distance and that orbits of any inclination are possible, and it is emphasized that there is a gradual transition from the realm of planets to the realm of comets.

The SIXTEENTH LETTER opens with an estimate of the velocity of stellar motion corresponding to a displacement of $\frac{1}{4}$ degree since Hipparchus. It is immediately recognized that such a displacement is the combined effect of the motion of the sun and of the star in question. The problem is further dependent on the respective position of the sun and of the stars in the local system of stars. In addition, there is the question of the center. Is it merely a geometrical point, or is it occupied by a celestial body of enormous density? Several reasons are given why the latter should be the case. One of them is the principle of analogy with the system of satellites and planets. The second half of the Letter is devoted to the conditions under which comets are visible and in particular a correlation is sought with the angle of inclination of cometary orbits.

In the SEVENTEENTH LETTER the first step is made toward giving a specific description of the fuller Copernican view of the universe. Our situation with respect to the common center of the local system of stars is compared with the task of astronomers on the moon to determine the center of the planetary system were the sun invisible to them. Then advance is made beyond the dark body in the center of the Milky Way to centers of higher systems and to the body at the center of the whole world-edifice. With an eye on Jupiter and Saturn and their satellites the dynamical consequences are investigated of the removal of those dark bodies from the center of their systems. Such a dark body is not supposed to have a retinue of planets and comets, but it must show phases. The second half of the Letter is a series of reflections on the topic discussed in the second half of the preceding Letter.

It is suggested in the EIGHTEENTH LETTER that the pale light in Orion might be the surface of a massive central body illuminated by suns orbiting around it. Are the variations observed in that light periodic, indicating a rotation? It is argued that without such a body in the center stars in the central region of their system would move very slowly in their orbit. Satellites, planets, suns, centers of systems of suns, centers of Milky Ways are bodies of first, second, third, fourth, and fifth rank respectively. If the central body of the world-edifice is a body of the thousandth rank, then the orbit of the moon becomes a cycloid of the 999th degree and that of the earth a cycloid of the 998th degree. Such is the fully Copernican view of the universe which implies strict subordination throughout. It is emphasized that the great number of comets around the sun balances the disturbing pull of Jupiter and Saturn. Connection is sought between the distribution of comets around the sun and of its orbiting in its system of stars. Finally, a similar connection is sought with regard to the zodiacal light, which is considered as the sun's atmosphere enveloping in part even the earth's orbit.

In the NINETEENTH LETTER distinction is made between the fully developed Copernican idiom and the ordinary one. Estimates are made of the respective sizes of cycloids superimposed upon one another as ripples are imposed on bigger waves. The world-edifice resembles a system of driving wheels connected through a thousand levels to the central wheel. This world picture implies that time and space are commensurate at each level. Fear about collisions is no longer justified. Confidence is expressed in spotting before long the dark central body, now called regent, of our system of stars and the regent of the Milky Way.

The elliptical orbits of planets are therefore shifted to ever higher ranking bodies as these are included, one after another, into the total picture. The process is illustrated by the properties of recurring series. Reflection on the zodiacal light forms the concluding part.

The TWENTIETH LETTER opens with the admission of the tentative character of such a theory of the world-edifice due to the absence of empirical verification. Then the whole theory is summed up in fifteen propositions, and in the case of each the measure of evidence is weighed. This is followed by an analysis of the method used in arriving at such a view of the world-edifice. The fundamental truth of mechanism is upheld, and it is suggested that gravity, which cannot be explained by vortices, might be reduced to pressure in the ether toward the center. The mechanical philosophy, of which experimental proofs are numerous, must be supplemented by cosmological (teleological) considerations. Finally, a comparison is made between the transitory character of terrestrial processes and the permanence of celestial bodies, a consequence of their perfect co-ordination into a planned, purposeful whole.

C) ITS COSMOLOGICAL BACKGROUND

This summary of the contents of the *Cosmologische Briefe* should make it obvious to anyone familiar with Kant's *Allgemeine Natur-geschichte*[53] that the two works are quite dissimilar in several aspects.[54] The principal point of similarity is the explanation of the Milky Way, but even there the differences are considerable. Since this explanation occupies only a small part of each work, their differences ought to appear very marked. To be sure, a hierarchical concept of the universe is advocated in both, and the idea of a massive body in the center of star systems and of systems of stars is also endorsed in both. But Kant's viewpoint is evolutionary, whereas that of Lambert is static. Kant advocates the infinity of the universe, a notion which is categorically rejected by Lambert. For Kant, galaxies, stars, planets and their systems are subject to a cyclic process of birth and decay repeating itself without end. For Lambert all major constituent parts of the universe keep their actual shape, structure, and correlation. In Kant's work the main body of discussion relates to the formation of planetary systems in which comets play only a very minor role. Lambert argues the anti-evolutionary view precisely with respect to comets. In his eyes they can in no sense be considered as immature or embryonic planets, and he builds heavily on the features of cometary orbits as contained in Halley's Table about which nothing is said by Kant. True, both Kant and Lambert are convinced about the presence of living creatures on all planets. Yet, while Lambert never gets involved in efforts to specify the physical and mental characteristics of denizens of other planets, Kant devotes the whole Third Part of his work to efforts of this kind. Even worse, Kant is convinced that the knowledge of gravity is enough to show why the denizens of Mercury are sluggish and why those on Saturn are all geniuses![55] Unlike Lambert, who is strictly Newtonian in the sense that he uses only the laws of motion in a central field of gravitation, Kant relies on forces of repulsion in addition to attraction (to say nothing of chance collisions, an anathema to Lambert). Clearly, the stereotype union of Kant and Lambert in histories of astronomy and cosmology has no unqualified support in their respective accounts of the universe.

The apparent similarities in Kant's and Lambert's cosmologies suggest, however, a common background which is not difficult to identify: Newtonian physics and Leibnizian teleology. The latter was conveyed to both Kant and Lambert by the works of Christian Wolff, who wrote voluminously and with complete confidence in teleology, of which he saw evidences in every nook and cranny of the physical realm.[56] In describing the denizens of other planets he offered in the style of a philosopher what Fontenelle had already made a common belief with his facile literary pen. Familiarity with Fontenelle and Wolff was only natural in the case of any educated man of those times. Trying to specify points of contact between Fontenelle and Lambert would be fruitless since it would amount to a discourse on the obvious and on the incidental. In quite a different a sense would it be fruitless to search for similarities between the *Cosmologische Briefe* and Wright's *An Original Theory or New Hypothesis of the Universe*,[57] to say nothing of Swedenborg's *Principia*.[58] These two works, especially the latter, were heavily steeped in a kind of mysticism wholly alien either to Lambert's piety or to his rationalism. The two *Principia* that had an obvious place in Lambert's thought were written by Newton and Descartes. About Descartes's thought Lambert had Malebranche for a source. At any rate, Cartesian vortices were still discussed by eminent mathematicians when Lambert was born[59] and they were still referred to when he was already dead.[60]

It is evident from the *Cosmologische Briefe* that Lambert was familiar with the collisional cosmologies of Burnet and Whiston, but this should not seem surprising. What should cause some surprise is that he made no mention of Buffon's theory of the evolution of the solar system.[61] It was the sensation of the late 1740's and early 1750's, and was meant to explain by one single collision of a comet with the sun the regularities of the planetary system, which Lambert emphasized time and again as incompatible with chance collisions.

In the *Cosmologische Briefe* no reference is made by Lambert to Chéseaux, although he uses repeatedly the model of the distribution of stars based on the close packing of spheres equal in size. Lambert's omission of Chéseaux might probably have been due to his unwillingness to repeat himself. By the time Lambert was seized with the idea of writing the *Cosmologische Briefe,* his *Photometria* was already in print. There Lambert discussed and rejected Chéseaux's solution of the optical paradox of an infinite universe of stars.[62] Lambert does not seem to have been familiar with Bentley's objection to the infinite universe on gravitational grounds,[63] nor with Halley's discussion in 1720 of the optical paradox.[64]

The most decisive background for the writing of the *Cosmologische Briefe* was the strong interest in comets during the middle part of the eighteenth century,[65] an interest heavily influenced by the appearance of the spectacular comet of 1744 and by the return of Halley's comet in 1759. But here, too, the analysis which Lambert gave of Halley's Table shows a thinker working in isolation, writing not even from books piled up on his desk, but solely relying on his memory. Some data emerged from that memory with clear reference to their original proponents, as is shown by Lambert's repeated mention of Derham; other data remained generic, and at times influenced only subconsciously the reasoning in the *Cosmologische Briefe.* In fact, Lambert later felt that he had forgotten to utilize Maupertuis and others as was already mentioned.

D)   AS VIEWED BY ITS AUTHOR

In late March and early April, after the printing of the *Cosmologische Briefe* had been completed, Lambert took advantage of the famed Leipzig Fair to have copies of it sent, together with his freshly printed book on the orbit of comets, to some of his friends. The letters accompanying the books give a priceless insight into his own way of looking at the *Cosmologische Briefe.* "I do not know," Lambert wrote on March 24, 1761, to Kaestner in Göttingen,

> "what you will think of the *Cosmologische Briefe.* You may very well amuse yourself with it. You will probably be surprised by the boldness with which I depict the structure of the universe and perhaps also by the abundance of all sorts of arguments which I submit there. I am extremely curious to see what luck this work is going to have with the readers and to cast its horoscope. As far as I can see, it is impossible to refute it solidly and perhaps the system [implied in it] will be dominant in the future and after some debates, but I do not believe that anyone would soon make a scrutiny of what I propose in some of these letters, because it demands long calculations and a series of observations."[66]

On the same day Lambert also wrote to Tobias Mayer, another acquaintance of his at the University of Göttingen. The letter[67] started with a reference to the Fair in Leipzig where Lambert hoped to find the means of sending his two newly printed works to Mayer as a token of his esteem. The one on the orbit of comets could cause no problem, "but I do not know what you will think of the *Cosmologische Briefe,* which are as daring as they can be." Equally daring on Lambert's part was his reference in the *Cosmologische Briefe* to Mayer as the one who provided conclusive evidence about the displacement of some stars. Lambert now admitted that all his assurance on this point amounted to his trusting a friend's words about Mayer's work. "I hope I was not misled by him," Lambert apologized to Mayer. "I know well that it is too late to gain assurance in the matter post factum. But your dissertation, unknown as it was to me, was too much to the point to be passed over in silence." With that Lambert came to the most revealing part of his letter: "But you will find indeed, Sir, that it was necessary to appeal to some experimental evidence in order not to demand of the readers to believe in proofs which are simply cosmological. These kinds of proofs are not yet fashionable and several who tried them out, have failed. It remains to be seen whether these letters have a better luck."

The "cosmological proofs" were, of course, a priori considerations about the universe, about its harmony and purpose. Their principal champions were Leibniz and Wolff, whose popularity was not at all high around 1760 even in Germany. Nor were "cosmological" considerations helped by the positivist view of science advocated by the French Enlightenment which was spreading across Germany under the influence of the Berlin Academy, the place which Lambert hoped to conquer with his rash of publications. But while his readers might have needed the a posteriori data, Lambert himself could have his certainty without them, or at least he seemed to suggest something very similar. In the same letter to Mayer he saw the principal source of future diffidence about the *Cosmologische Briefe* in the fact that "we are not yet sufficiently Copernican. But at the present," he added with his deeply ingrained self-assurance, "I do not believe that one can become more Copernican than I have become in these letters."[68]

There was also the same touch of self-esteem in the remark which he added to his report that since his sojourn in the Netherlands he had hardly done anything except work on his publications. He hoped to complete them soon, to have among other things time to do the computations and observations suggested in the *Cosmologische Briefe* for the detection of "what I call the center of our system of fixed stars." The center was one of those dark massive stars which Lambert called regents and about which he now remarked: "I do not believe that anyone else would soon be disposed to look for them."[69] Neither Mayer, nor anyone else, is known to have looked for those dark celestial bodies, or to have engaged in the calculations proposed by Lambert.

III   The *Cosmologische Briefe* and Posterity

A)   ITS EARLY RECEPTION

The failure of contemporary astronomers to react to Lambert's suggestions was symbolic of the reception which was in store for the *Cosmologische Briefe*. To begin with Mayer—he did not do what Lambert begged him to do, namely, to express his "frank sentiments" about the work. Nor did Kaestner, who must have had some role in the appearance of a brief review in the September 19, 1761 issue of the *Göttingische Anzeigen von gelehrten Sachen*.[70] Much of the review dealt with Lambert's arguing on behalf of the very great number of comets. What Lambert said about the Milky Way was recalled in the bare remark that the Milky Way was, according to him, the ecliptic of stars. Nothing was said about the ever more inclusive star systems and the ever higher ranking regents. One wonders if the readers of that report had found convincing the concluding remark: "Lambert develops with much ingenuity this and several new thoughts which we must,. however, leave to be read in his own words."[71]

Lambert seemed to have received in due time a copy of the review, but three years later he was still puzzled by Kaestner's failure to acknowledge the book which, as Lambert put it, must have reached him or else the review of it could hardly have appeared.[72] In general, Lambert did not solicit the views of prominent men of science on the *Cosmologische Briefe*. The classic case was Euler, the third of those to whom Lambert wrote on that March 24, 1761. Neither in this letter,[73] nor in the ones which Lambert wrote to Euler in February and in June,[74] is there a reference to the *Cosmologische Briefe*. All these letters served the purpose of gaining Euler's support for Lambert's candidacy for membership in the Berlin Academy. Lambert seemed to realize that the road to its Class of Science had to be paved by works different in character from the *Cosmologische Briefe*. Indeed, in Lambert's correspondence between 1761 and 1764 with Sulzer, who was his chief champion in the Berlin Academy, and who was determined not to let the Russian Academy secure Lambert's talents for itself, it was not the *Cosmologische Briefe*, but the *Photometria* and the works on comets and on the path of light rays that were mentioned.[75] The same pattern was evident in Lambert's letter of April 8, 1761 to d'Alembert,[76] a fact certainly revealing in view of the eagerness of any author to speak of his freshly printed book.

Lambert's most effusive communication about the *Cosmologische Briefe* which came in the wake of its publication was not addressed to a man of science, but to his former student, Baptista von Salis. "The *Cosmologische Briefe,* of which I have the honor of offering you the copy here attached," Lambert wrote on April 10, 1761, "are written in a manner that each reader will think of it differently and I could not do it otherwise. I had to say myself all that the most incredulous will find fault with, and he who is not entirely in the position of passing judgment on it, will take it in a stride. But there will be others, who will weigh my arguments and who will push them further through the observation and computations which I propose. I have no reason to doubt that gradually they will succeed, but until then I had to set aside the positive tone. This is also the reason for the long preface and for the last letter. The trouble is that there are very few competent judges. What is at the basis of the matter is not within the resources of everyone, as regards the proofs. He who does not sufficiently grasp their weight can easily treat the rest as an astronomical novel, and I am the first to suggest to them this idea and a number of others, which they can form themselves. Meanwhile, the idea which I give there of the universe is complete, and I have found no way of seeing that it would be possible in another way as far as the order, connection, and harmony of the whole are concerned."[77]

Four months later, on October 1, Lambert wrote again to Baptista, and the *Cosmologische Briefe* was the chief topic.[78] Lambert referred to his plans of suggesting to Professor Bodmer, a well known poet in Zurich,[79] to utilize the vistas of the *Cosmologische Briefe,* since he himself had mentioned in Letter XVII that it would take a poet to do full justice to the harmony and arrangement in the universe. At this point let it be recalled only what he thought about the philosophical and scientific merit of the work:

"In the philosophical world one will rank these *Letters* slightly below the *Theodicy* of Leibniz, and I would in vain continue writing astronomical novels, people would not believe me though they would demand that I go on with them, and they would firmly claim that I must still have a number of manuscripts in this style, and that I am obligated to publish them. But one must see what the astronomers will say. The dice is cast, the sails are set."[80]

The astronomers said nothing and hardly a breeze came. A very recent statement, that the publication of the *Cosmologische Briefe* created a great sensation,[81] has no foundation in the records. Sometime in late January, 1762, a letter of J. Wegelin to Bodmer was forwarded by the latter to Lambert possibly because Wegelin wrote that "Lambert's [cosmological] system really provides the material for an admiring adoration of the Divine Being."[82] Wegelin, professor of history in Berlin and member of the Academy of Sciences, was hardly the person to evaluate the scientific merits of Lambert's cosmology. Also, the rest of his letter was not at all an unqualified praise of Lambert's work. Its a priorism was the implicit target of Wegelin's doubt whether there could ever be a satisfactory correspondence between a priori postulates and the actual structure of the universe. Lambert was not excepted as Wegelin remarked: "All physicists and astronomers have been until now but commentators of the universe trying to decipher the infinite dark details from a few clear ones. They laid down as starting point a few observations and added the rest from their own heads."[83] The general idea of perfection could not be equated with any physical theorem and, Wegelin added, Bernoulli's idea of vortices was just as susceptible of being applied to the entire universe. Tellingly, Wegelin's concluding remarks directed attention to Bacon's empirical program.[84]

Lambert copied Wegelin's letter and added to it some comments which he possibly intended to send to Bodmer and through him perhaps to Wegelin. Lambert, looking with eager eyes at the Berlin Academy, could only be flattered that one of its members was commenting on his work. Yet, had Lambert already been on equal rank with Wegelin, it is unlikely that he would have spoken of Wegelin as a "great expert" who finds his work "worthy of attention and tries to find the gaps still present there and shed light on them."[85] That Wegelin was not really an expert in Lambert's eyes came through clearly in the remainder of his comments. He seemed to lecture Wegelin who touched upon the proper use of analogy in cosmology, a point all the more crucial because Lambert's cosmology was based on the postulate that all levels of the universe reflected the analogous realization of the same pattern. The genuine scientific strain in Lambert's thinking asserted itself as he noted: "The analogy is not sufficient for a secure determination of truth, but it gives the opportunity for guessing it and for establishing it fully with the help of experiments and investigations. And herein lies the difference between use and misuse."[86] To this general rule he added references to a dozen or so pages of the *Cosmologische Briefe* where he warned against an arbitrary use of analogy and specified the observational work to be done to secure complete validity to his speculative conclusions.

The next year the *Cosmologische Briefe* found favorable mention in Kant's essay, "On the Only Possible Ground for Proving of the Existence of God," in which he gave a detailed summary of his own cosmogony as an illustration of the argument from design.[87] As he touched in the Preface on this point Kant regretfully noted that the full form of his cosmogony, the *Allgemeine Naturgeschichte,* published anonymously in 1755, came to the notice of but a few. It remained unknown, Kant continued, even to "the famous J. H. Lambert who six years later set forth in his *Cosmologische Briefe* in 1761 exactly that theory about the systematic arrangement of the universe, of the Milky Way, of nebulous stars and so forth, which one finds in the First Part and similarly in the

Preface of my foregoing work." What Kant now added revealed something of the deferential attitude of a still struggling academic towards an already famous man, something of his self-consciousness, and something of his readiness to take lightly the enormous differences between his and Lambert's cosmology: "The agreement of the ideas of that insightful man with those which I had submitted at that time, and which correspond to one another even almost to the smallest details, increases my confidence that the [cosmological] system will obtain more confirmation in the future."[88] But the publication of Kant's essay and its three more editions within the next thirty years[89] drew attention neither to Kant's cosmogony, nor to Lambert's work.

It was from another direction that, toward the end of 1763, the wind seemed to find the sails of the *Cosmologische Briefe*. In the December issue of the *Nouvelliste Suisse* there appeared a brief notice about the *Cosmologische Briefe* and about the plan of publishing in the subsequent issues a French translation of it.[90] As a justification of the translation it was mentioned that among scientists connected with the *Journal encyclopédique,* the leading French organ of scientific and literary criticism, there was considerable enthusiasm for the *Cosmologische Briefe*. The anonymous translator first, however, gave in the January 1764 issue a brief and elementary explanation of the solar system under the title, "Général du Sistème [sic] de l'Univers."[91] The February issue carried the translation of the first two Letters, and with that the wind died down.[92]

More than a year passed before enthusiasm turned into printed words in the *Journal encyclopédique*. In the June 1, and 15, 1765, issues there appeared a detailed and careful summary of the *Cosmologische Briefe*.[93] The competent, anonymous author of the summary did not suggest that Lambert offered anything spectacularly new with his explanation of the Milky Way.[94] While it was admitted that one could not deny that "the work is not full of new, profound, and sublime views," the concluding remark conveyed subtle scepticism: "even if this work is but a philosophical novel, it is at least the most beautiful and most ingenious of such novels."[95]

Lambert's appointment to the Academy came about with no reference whatever to the *Cosmologische Briefe,* as can be seen in Lambert's letter of February 17, 1765 to Frederick the Great, in which he spoke of his achievements in philosophy and physics. Cosmology was not mentioned.[96] It was only two years later that Lambert proudly stated to the king that "the public considered my works as second volumes to be attached to the one on optics by Newton, to the one on the plurality of worlds by Fontenelle, and to the "Organons of Aristotle and Bacon, as well as to the work of Locke."[97] The foregoing reaction of the public to the *Cosmologische Briefe* still had to evidence itself publicly. Lambert's

Neues Organon certainly created a greater interest than his *Cosmologische Briefe*. It was in reference to the former that J. J. Simler of Zurich concluded his letter of April 6, 1764 to Lambert, "May you live long, second Leibniz!"[98]

As to Kant, part of his interest in the *Cosmologische Briefe* centered on his vindicating his own priority with respect to the explanation of the Milky Way. In his letter of November 13, 1765 to Kant, Lambert insisted on his unfamiliarity with Kant's work at the time of the composition of the *Cosmologische Briefe*. Lambert now also revealed that about the time of its printing he had heard of somebody in Nürnberg planning a German translation of a work by Thomas Wright of Durham, which contained ideas similar to his on the Milky Way. The same letter of Lambert also evidenced his resolve not to yield to an uncritical a priorism. He chided students of Greek literature who, upon reading the a posteriori explanation of the Milky Way in the *Cosmologische Briefe,* would leave no stones unturned to show that the ancient Greek sages had said the same long before and on a purely a priori basis.[99]

From the next year on there developed a fairly regular correspondence between Lambert and George Louis Le Sage, the scientist-philosopher of Geneva, who is today remembered only for his ingenious though faulty explanation of gravity by the impact of "ultramundane" (ethereal) particles on ordinary matter.[100] A minor figure in physical science even in his time, Le Sage was the only scientist to take, while Lambert still alive, an active interest in the *Cosmologische Briefe*. In a letter probably written on August 23, 1767, Le Sage mentioned to Lambert his plans to have the work translated into French. In addition he wanted to have the translation printed with his "teleological" notes.[101] In his letter of April 14, 1768 Lambert mentioned to Le Sage that years earlier he had begun a teleological treatise in the style of the *Cosmologische Briefe* on terrestrial processes. "But the topics became so bristling in the first letters that I saw that they had to be restated or that the letter-form had to be abandoned until I could see that the subject matter admitted a more positive and philosophical tone and the texture of reasoning could be more cogent."[102] One wonders if the same could not be said about the *Cosmologische Briefe*.

Lambert's next and long letter of October 1, 1768 to Le Sage[103] is a goldmine of information about the *Cosmologische Briefe,* although its main topic was Le Sage's theory of gravity. After noting that on the basis of Le Sage's theory "more than eternity was needed to produce a planet," Lambert recalled his having stated in the *Cosmologische Briefe* that a satellite always had been a satellite and never a disoriented comet.[104] Clearly, evolution on a cosmic scale was not to Lambert's liking. Lambert now registered his pleasure about Le Sage's efforts to secure the translation, and quickly offered his rules on the art of making a good translation. The rules revealed something of Lambert's awareness of the not-so-lucid character of his own style. He seemed to blame it on the German idiom, but Lambert himself was to be blamed for ignoring time and again rules of German syntax in the *Cosmologische Briefe.* "It is true," he wrote to Le Sage, "that the German language, as well as the Latin and the Greek, allow somewhat long phrases, whereas in French one does not know how to make them. Very often the translators enervate them by cutting them and especially by omitting the *conjunctions,* and thus the connection is lost." After this it could hardly appear convincing what Lambert added in the next breath: "It is true, on the other hand, that they can be cut without risk."[105]

In the same letter Lambert disclosed a new aspect of his pleasure in having written the *Cosmologische Briefe.* "I am ready to believe that most of the ideas contained in the *Cosmologische Briefe* can be found dispersed here and there in my other works. I would not have omitted writing these Letters as it always seemed to be a good idea to make a whole of those thoughts."[106] Equally revealing was Lambert's remark that he had already been shown a copy of Kant's cosmology but that he could not obtain for himself a copy "of that rather unknown work of Kant." As to Wright, it was still only by hearsay that Lambert knew of his work: "An Englishman, I believe Mr. Wright of Durham, must have also composed a treatise on the system of fixed stars. I am told that he is rather backward."[107]

Lambert's cosmology was only indirectly involved in what should seem to be the most remarkable part of his letter to Le Sage when taken in the philosophical context of the time. Lambert made it clear that while his ideal in philosophy was the strict logic of mathematical reasoning, it was not to be forgotten that when any object was submitted to the rigor of quantitative analysis this made it appear in a wholly different light. He certainly did not wish to become a reductionist and even went to the defense of the "occult qualities" used by the scholastics. Only laziness to study the scholastics, he noted, could lead to the misconception that more than names became changed through criticism and dismissal of scholastic philosophy.[108]

Toward the end of the same year Lambert mentioned again, in a letter to George Holland, his abandoned project to write a terrestrial or "sublunary" sequel to the *Cosmologische Briefe.*[109] In the same letter Lambert referred to Charles Bonnet who, as Lambert put it, "brought out many beautiful details about the purposeful arrangement of nature."[110] Bonnet, who leaned toward the notion of evolution, could not be to Lambert's liking. Bonnet, he remarked, was not strong enough in applying the analytical method. At any rate, Lambert added, "the laws of change can easier be found than the laws of structure."[111] Such was a surprising claim but not on the part of a convinced anti-evolutionist.

In his letter of December 16, 1769 to Carl R. Hausen, professor of philosophy in Halle, Lambert made reference to the "few letters" translated in the *Journal helvétique.*[112] Johann Bernoulli later remarked that the publication of those "few" letters played a part in thwarting the effort of Le Sage on behalf of a complete French translation. A former student of his had reached mid-way in the project when it failed.[113] The next year brought some luck for the *Cosmologische Briefe* when Johann Bernhard Merian, Swiss director of the Literary Class of the Academy in Berlin, came out with a French summary of the work under the title, *Système du monde.*[114] It was less than one half of the original in length and was written in a plain, expository style to the complete exclusion of the letter-form. Instead of twenty letters there were now two Parts, each with nine chapters. With respect to contents the condensation contained all the major ideas and arguments in the original but in the whole it was more Merian than Lambert. It was published in Bouillon where the *Journal encyclopédique* also had its editorial office and printing shop. Merian's summary was reviewed in the lead article of the November 1, 1770, issue of the *Journal,* but in a manner which could hardly please Lambert. His "most profound astronomical and geometrical erudition," the reviewer wrote, was now matched by that "accomplishment in style and finesse in eloquence which so markedly characterize Mr. Merian." This followed the reviewer's remark that "Lambert's profound, learned letters would be almost completely ignored in France, where probably nobody would have heard of them, had we not seen to inserting an extract [of it] in one of the 1765 issues of our Journal."[115] Clearly Merian gained more than Lambert. The latter's profit was another flurry of interest in the original, but again enthusiasm did not go beyond the confines of private correspondence.

During 1771 several letters were exchanged between Lambert and Giuseppe Toaldo of Padua, who read the review in the *Journal,* and subsequently the condensation by Merian. Toaldo was enthusiastic. He characterized the condensation as the most beautiful text written on the plurality of worlds. The condensation, which he considered as giving only the "popular sense" of the original, whetted his appetite for a complete French translation.[116] In his letters to Toaldo, Lambert expressed his satisfaction with Merian's condensation, resented the reviewer for having given too much credit to Merian, and voiced his hope that an Italian translation might eventually be forthcoming.[117] Lambert's dissatisfaction with the reviewer appeared also in his letter of October 6, 1771, to Le Sage.[118] It is doubtful that he had learned of Bonnet's enthusiasm about the condensation by Merian. In a letter to Haller, Bonnet took the view that there was now no more need for an astronomer to interpret the original.[119] This remark of Bonnet hardly threw a favorable light on his familiarity with astronomy which the *Cosmologische Briefe* offered, on the whole, on a sufficiently simplified level. Interpreters were needed to make sense of Lambert's phrases on more than one occasion. About those phrases Felice Fontana, professor of mathematics at Padua, wrote upon receiving Lambert's German books to Kaestner in 1773 that they were often difficult to understand because of their length. To illustrate his point Fontana quoted Pope on Dryden: "The copious Dryden wanted, or forgot,—The last and greatest art, the art of blot." Still, Fontana insisted that Lambert was a genius, and therefore the faults of his style had to be overlooked.[120] His subsequent correspondence with Lambert contained no reference to cosmology!

Earlier that year Lambert learned from Boeckmann that he was instructing in a "Socratic fashion" his princely patron about the contents of the *Cosmologische Briefe* and that the Prince could not have absorbed its ideas with greater admiration.[121] Such was hardly a comfort to Lambert who, in the same year, could read Lalande both praise and damn the *Cosmologische Briefe.* Lalande, eager to dissociate himself from rumors that made him predict the imminent collision of the earth with a comet, readily endorsed Lambert, "the well known mathematician," who saw no possibility for it in his *Cosmologische Briefe,* "a work full of imagination, insight, and knowledge." But toward the end of his brochure Lalande characterized Lambert's inferences to a very large number of comets as "very vague and very chancy, although ingenious and learned."[122] Clearly, the latter adjectives could not have appeared too convincing in the close company of the former.

Lalande's brochure was soon followed by a massive analysis of the possibility of the earth colliding with a comet. Achille Pierre Dionis du Séjour's *Essai,* "on comets in general and especially on those that can approach the earth,"[123] could only be read by an expert in celestial dynamics such as Lambert; in its more than three hundred pages some fifty comets were subjected to a quantitative analysis to see whether the possibility of collision with the earth was a serious one. The topic closely touched the heart of the *Cosmologische Briefe* since case after case showed the extreme remoteness of a possible collision. But Dionis du Séjour failed to mention even Lambert's book on computing the orbit of comets, let alone his cosmological work, although he listed half a dozen titles relating to cometary orbits.[124] Even worse, for Lambert at least, was that he was not mentioned when Dionis du Séjour discussed the consequences of removing Jupiter and Saturn from the midst of their satellites.[125] Was not this topic given special attention in Lambert's work, and did he not refer again and again to the impossibility of viewing the moon as a comet captured by the earth, a topic also treated by Dionis du Séjour in a detailed form?[126]

Among those passing judgment on Dionis du Séjour's book on behalf of the Académie was Lalande.[127] His already printed and peculiar praise for the *Cosmologische Briefe* can easily explain why Lambert's subsequent letter to him contained no reference to cosmology.[128] Nor, for that matter was cosmology mentioned in a letter which Lambert wrote about the same time to Maupertuis,[129] well known for his interest in comets, or in the last pieces of Lambert's correspondence with Euler whose popular exposition of the study of the physical universe was the sensation of the 1770's.[130] Lambert's reluctance to take up with first-rate "geometers" the subject matter of his *Cosmologische Briefe* may suggest a growing awareness on his part that his work was too speculative to suit the taste of the official science of the day.

Praise for the *Cosmologische Briefe* came around that time from the pen of secondary figures, and Lambert's references to it were also conveyed to such. In 1775 Jean Trembley of Geneva informed Lambert of his enthrallment while reading through Merian's condensation of Lambert's cosmological ideas.[131] Lambert's last reference to the *Cosmologische Briefe* seems to be in his letter of May 31, 1777, to Aloysius Havichorst, professor of metaphysics at the University of Münster. It was a complaint about the neglected status of cosmology: "Considered a priori, it seems to me to be the poorest part of metaphysics. Taken a posteriori, physics and astronomy provide rich material for it . . . I have given in the *Cosmologische Briefe* a few ideas about the greatness of God. Even [Saint Augustine's] *De civitate Dei* can receive from it some clarification."[132] Clearly, Lambert looked for more light than could be provided by the Enlightenment. Less than three months later he passed into eternity.

That the reply, which Havichorst sent to Lambert on July 14, contained no reference to cosmology,[133] let alone to the *Cosmologische Briefe,* was symbolic of what was to come when the deceased was eulogized in word and print. One of the first to publish a eulogy was Johann Elert von Bode, famed author of a popular astronomy,[134] the future Astronomer Royal in Berlin, and already Lambert's successor in charge of the publication of a yearly astronomical ephemerides. When Lambert died, the volume of the *Astronomisches Jahrbuch* for 1780 was already in press, but Bode succeeded in adding a last-minute notice on Lambert.[135] It was silent on Lambert's cosmology, in strange contrast to Bode's glowing though brief and generic endorsement of Lambert and Kant as cosmologists in his two-volume manual of astronomy.[136]

Formey, whose duty it was to eulogize Lambert in the Berlin Academy, referred to Lambert's "sublime cosmological ideas." Yet, while he spoke in some detail of all of Lambert's books, the *Cosmologische Briefe* was not mentioned.[137] Cosmology was overlooked in the *Teutscher Merkur's* lengthy appraisal of Lambert's character and work in its September 1778 issue.[138] A year later G. F. Lichtenberg, the leading scientist in Göttingen, published anonymously a brief biography of Lambert. Its brief reference to the *Cosmologische Briefe* and to Lambert's "instructive and likely speculations about the arrangement of the system of fixed stars,"[139] contrasted sharply with the effusiveness of enthusiasm with which Lichtenberg greeted in 1785 Herschel's second account of his observations and conclusions on the construction of the heavens.[140] At any rate, Lichtenberg had already remarked in a letter to Herschel that Lambert's ideas were but vague anticipations of Herschel's discoveries.[141]

Sometime before Herschel's startling disclosures about the realm of nebulae brought discredit to Lambert's neat schematism of a hierarchically ordered universe, Lambertian cosmology had suffered another blow in another, equally sensitive domain—the realm of comets. The blow was all the sharper because in Alexandre Pingré's *Cométographie,* a massive historical and theoretical study of comets,[142] a prominent role was given to Lambert's method of computing cometary orbits.[143] But Lambert was ignored when Pingré discussed the nature and properties of comets,[144] their atmosphere,[145]—topics very relevant to the main line of argument in the *Cosmologische Briefe.* Lambert was not mentioned when Pingré took up the question of the ways of estimating the number of comets![146] In view of the generally shared belief in the existence of denizens on each and every comet, Pingré could naturally ignore Lambert in this connection, but his discussion of the topic showed a measure of shortsightedness exceeding that of Lambert. In discussing Buffon's theory of planetary evolution based on a comet's collision with the sun, Pingré set great store by its elementary shortcoming in celestial mechanics.[147] It did not occur to Pingré that Buffon's theory was equally defenseless against the fact, emphasized by Pingré himself, that the mass, which one could attribute to comets on the basis of celestial mechanics, was exceedingly small compared with even the mass of planets, let alone with that of the sun. It was all too clear to Pingré that the absence of perturbations which comets were expected to produce was an incontrovertible proof of their exceedingly small mass. Since that small mass was distributed in a relatively large volume, it followed that the average density of comets was far below that of ordinary vapor. Inference of this kind must have been in the back of Pingré's mind when he emphasized that denizens of comets must have had extraordinary physical characteristics.[148] But did it then make sense to speak of them, and most categorically so, if they resembled angels rather than corporeal intelligent beings? If such plain logic escaped the "geometers," then it was certainly not to be noticed by others, especially when already under the sway of Lambert's vistas.

30

Indeed, no trace of plain logic was apparent in the sole unqualified outburst of enthusiasm about Lambert's cosmology to appear in print. It came from Bonnet, the famed naturalist, who had already admitted in private communication his inadequacy to cope with the astronomy of the *Cosmologische Briefe*.[149] In a lengthy note added to the revised edition of his *Contemplation de la nature*[150] Bonnet advised his readers (and they were very numerous) that, "in order to acquire the highermost ideas of the extent and population of the universe, one must read and meditate upon the admirable *Système du monde* of the profound Lambert, a work which one would believe to have been written by a celestial intelligence rather than by an inhabitant of the earth."[151] Bonnet's rapture reached its peak as he referred to that central body around which myriads of stars, planets and comets revolved, a body "which in turn is dominated by a central body even more powerful of which it is but a satellite . . . What a satellite! . . . Here the mind loses its ability to admire and the astonishment turns into stupor: how such a spectacle offers itself to the eyes of a single mortal! This mortal was therefore an angel, disguised under the form of a human, or had he been carried to the Third Heaven?"[152] Then, after having described the innermost central body in the universe, Bonnet could only give of Lambert's achievement as an interpreter of nature an estimate that made one forgetful of Newton, Leibniz, and of many others: "It has become revealed to us in these last times that the universe is really an immense work of mechanism, composed of an innumerable multitude of pieces of different size and density, which, engaged into one another, or tied to one another by a general law, are joined by the same law to a master wheel, to a prime mover, whose inconceivable activity penetrates from body to body, from big to small, across myriads of spheres, to the most remote extremities of the universe."[153]

That the universe was an interconnected mechanism was hardly news by Lambert's time, nor were his claims, for which Bonnet also gave him undue credit, about its being full of living beings, all witnesses of the rich variety of the purpose of creation. For all of Bonnet's superlatives about Lambert, the real gain seemed to devolve on Merian's *Système du monde*. Bonnet's reference to it might have been instrumental for its coming out in a second edition only three years later, in 1784.[154] With this new edition the *Système du monde* not only eclipsed the original, but also blocked the publication of its almost complete French translation by Antoine Darquier, astronomer of Toulouse, although Lalande himself stood behind the project. By the time Darquier's translation saw print, in 1801, through the good services of J. M. C. van Utenhove of Utrecht, under the appropriate title, *Lettres cosmologiques sur l'organisation de l'univers,*[155] Merian's condensation scored two more successes. In 1797 Mikhail Rozin, a very minor figure, published in Saint Petersburg a Russian translation of the condensation of which only some of the largest Russian libraries have now a copy.[156] Three years later a certain James Jacque brought out in London Merian's condensation in English translation which today is even rarer than the very rare original.[157] This English version was further excerpted in 1952. It is that excerpt[158] of the English translation of the French condensation of the German original which for the past quarter of a century served as the principal source of information for the English speaking public (and even for most cosmologists!) about the contents of the *Cosmologische Briefe.*

Clearly, with condensations dominating the scene, no serious study was to appear on Lambert, the cosmologist. Darquier's translation and Utenhove's notes represent a lonely effort to do scholarly justice to the *Cosmologische Briefe* as a document of cosmological science. Perhaps they did too much justice to it! Not that either of them, and especially Utenhove, had neglected to report the latest and most important developments. These were, above all, Laplace's work on the perturbation of planets and Herschel's advance into the realm of nebulae. In his numerous references to both, Utenhove did not try, however, to convey the true extent to which they went counter to Lambert's hopes. Laplace's work poured cold water on Lambert's reasoning that in order to explain some puzzles of the planetary system one had to look for a faraway dark body ruling our system of stars. Laplace also disclosed a feature of very massive stars which deprived Lambert's dark bodies of the possibility to be detected optically. In addition, Laplace seemed to find explanation in an evolutionary picture of the solar system for some of its features which Lambert singled out as possible clues to his own ideas. Worse even, the realm of stars, the Milky Way, and other galaxies revealed themselves through Herschel's telescopes in a manner very different from what Lambert expected from "further experiences." Nor was it spelled out by Darquier and Utenhove that the new developments in the count and statistics of comets, which seemed to bring a stunning measure of success to at least part of Lambert's cosmology, were not achieved with an eye on him. Darquier and Utenhove were alone in exploiting on behalf of Lambert the updated list of comets. It was compiled by F. von Zach from several fragmentary lists drawn up by E. Prosperin, astronomer of Upsala, and published as an appendix to Olbers' booklet on computing the orbit of comets in 1797.[159] But von Zach, who took upon himself the publication of Olbers' manuscript, mentioned Lambert only in passing[160] as one who excluded the possibility of collisions among celestial bodies in his *Cosmologische Briefe.* Neither von Zach nor Olbers had sympathy for Lambert's cosmology, a point never intimated by Utenhove.

The case was similar with the first major publication of J. H. Schröter, the German Herschel. It was an analysis of his observations of Venus with his 27-foot telescope,[161] the wonder of Lilienthal. The observations were followed by a report on the great nebula in Orion, illustrated with a large engraving of its shape and varying hue.[162] Had Lambert lived, his heart would have jumped on reading Schröter's list of changes in that nebula. Did not Lambert stake so much on those changes? But Schröter, a point not mentioned by Utenhove, kept silent on Lambert's cosmology, although in one respect, the fulness of the world with inhabitants on each and every globe, he was even more Lambertian than Lambert, if this was possible. The almost complete lack of atmosphere on the moon, its surface pocked with craters, the heavy atmosphere of Venus (Schröter believed he could see many mountains beneath it!) were no arguments against the principle of purposive plenitude. "Enough!" he wrote in a way of conclusion,

> the whole nature speaks on behalf of what has been repeatedly asserted, namely, that each celestial body can be so constructed by the divine Wisdom that it might be provided with living creatures, constituted according to its physical frame and destined to praise God's might and goodness, and that the infinite greatness of the Creator ought to be manifested just as well in the analogous manifoldness of the physical conditions of celestial bodies as in the almost infinite diversity of creatures.[163]

In length, complexity, and content, the phrase was certainly Lambertian and even cosmological, but only in a sense which was already outmoded when Schröter wrote. The change in mode had much less to do with the ever shifting moods in philosophy than with cosmology which was about to become a science in the modern sense.

B) ITS RELATION TO MODERN COSMOLOGY

Already in the early nineteenth century the original, its condensations, and especially its French translation were on their way to becoming rarities, a process expressive of the general disinterest in Lambert's cosmology. The principal reason for this was undoubtedly the overwhelming impression made by Herschel with his telescopic discoveries which ushered in modern cosmology. Some contemporaries of Herschel complained that his discoveries had to be taken on faith, since he alone had those giant telescopes.[164] Yet the number of competent people whom Herschel entertained to view the heavens with his gigantic instruments was too great to justify any scepticism about what he had reported. Moreover, his reports exuded an unmistakable factuality. He not only described what he saw, but was very careful in recording the positions of double stars, clusters, and "nebulous stars" that came within his view in ever larger numbers and variety. Estimating the distances of this unexpected host of celestial objects was another matter. At least Herschel presented very clearly his quantitative procedure.[165] On the basis of his data there could be no possibility for doubt about the cause of the visual appearance of the Milky Way. The Milky Way could only be thought of as a host of stars confined within a disk-shaped space with rather uneven contours. In arriving at this conclusion in 1784 he might have been influenced by Thomas Wright,[166] but he was in all appearance unaware of the works of Kant and Lambert. Kant's work had largely remained unknown even for Germans for another decade, and Herschel had already been four years in England when Lambert's book was published. Herschel made the wise decision of becoming as much as possible a part of his adopted country. From almost the moment he had landed in England he started instructing himself from English texts in astronomy. At any rate there was nothing comparable in German to works by Smith and Ferguson.[167]

But the same works were not about cosmology for which Herschel's papers contained basic building blocks. Even more astonishingly, those building blocks were at a first look the same ones about which Lambert had wistfully speculated a generation earlier and saw in their eventual verification by observation the final triumph of his cosmology. The fulness of the world as postulated by Lambert demanded more planets and satellites. It was with the discovery of Uranus, the first new planet in recorded history, that Herschel burst into the scene twenty years after the publication of the *Cosmologische Briefe*. It was again Herschel who first sighted satellites around the new planet and also some new satellites around Saturn. The motion of stars was a primary requisite in Lambert's cosmology, but for evidence he could only refer to the report that comparison of ancient and modern star catalogues indicated a displacement in the position of several stars. Herschel provided evidence that stars in the area of Hercules appeared to move away from one another and this meant that our sun was indeed moving in that direction. Again, the determination of the distance of fixed stars was a pivotal demand in the cosmology of Lambert, but here, too, it was Herschel who first offered reliable estimates. As to the Milky Way, its explanation as a layer of stars could only be a most plausible tenet until Herschel started gauging the heavens.[168]

One could indeed be tempted to say that Herschel took his research topics from Lambert's *Cosmologische Briefe*. But Herschel had long made all his major discoveries and had already formulated all his major speculations when he made his sole reference to Lambert's cosmology which he found "full of the most fantastic imaginations."[169] Apart from that it can easily be seen that the similiarities between Lambert's and Herschel's programs are superficial. A small but telling evidence of this is their respective thoughts on double stars. For Lambert their motion around a common center of gravity showed that while two stars still could form a stable and orderly system without a third star at the center of their orbits, such was not a possibility with a large number of stars. They needed a regent and a "despotic" one at that.[170] Herschel observed the motion of binary stars, and felt satisfied on finding that their paths were in full agreement with Newton's law of gravity.[171] While the universal cosmic validity of that law was for Lambert a pointer to the existence of regents, it did not occur to Herschel to look for them or to speculate on them even when he saw by the score magnificent globular clusters with myriads of stars in each. While Lambert might have seen a deadly blow to his uniform systematization of the heavens in the rich variety of nebulae spotted by Herschel, the latter took them as they were. The rigid permanence of the shape and correlation of celestial bodies as advocated by Lambert was a far cry from Herschel's willingness to look at the world of nebulae as a luxuriant garden in which ever new forms blossomed.[172] The rich variety of the universe was for Herschel a proof of the Creator, whereas Lambert argued the necessary fulness of the universe from considerations about the Creator. Most importantly, unlike Lambert, Herschel did not let this belief of his intrude into his actual discourse on the construction of the heavens. To try to specify the intents and plans in which the infinite wisdom evidenced itself in the created realm was never a temptation for Herschel. Last but not least, while Herschel made passing references to denizens on other planets,[173] he never tried to make astronomers of them.

Indeed the differences between the cosmological speculations of Lambert and the cosmological observations of Herschel could not be greater. The difference was not simply that one was speculative and the other observational. Such a categorization would fit the cubbyholes of positivists, but it would do no justice to the largeness of mind of both Herschel and Lambert. While Lambert at times could sound a priori and in a most complacent manner, he had always a genuine interest in precision instruments[174] and a great concern for future astronomical observations. While Herschel spent much of his time polishing his mirrors and observing the sky, he emphasized that of two errors, of speculating too little or too much, he preferred to be guilty of the latter.[175] The real difference between Lambert and Herschel, the cosmologists, has to do with the fact that science is neither pure speculation nor sheer experimentation but a creative interplay of both. Pivotal as is the role of the speculative mind, its eyes must be supplemented with instruments to secure the breakthroughs provided by the mental vision. It is this lesson which comes forth with an almost brutal force in a comparison of Lambert and Herschel.

Herschel's achievements might not appear in their true greatness in an age which has long grown accustomed to seeing galaxies as the building blocks of the universe. The true measure of Herschel's greatness can only be gauged by seeing him against his own cosmological background. The most telling part of that background is formed by Lambert's *Cosmologische Briefe*. There one can see a great mind ready to soar into the farthest reaches but not equipped with a reliable guiding mechanism and unable to form any idea about the true measure of the distance covered on the wings of his speculations. The background provided by Lambert shows the greatness of mind handicapped by the absence of an effective ability to extend the realm of observation. It is the same background which brings into powerful relief the fortunes of an equally great mind which was not deprived of that ability.

The striking difference between the cosmological literature of the eighteenth and nineteenth centuries is indeed the difference between the absence and presence of that ability. Whereas cosmological ideas were not frowned upon, after Herschel's major improvements in the art of constructing telescopes the observational part of cosmological work was to dominate. At any rate, Lambert's cosmological ideas were not widely spoken of as the nineteenth century began its course. Only a few specialists and bibliographers were to look up Lalande's *Bibliographie astronomique* to find there buried among thousands of titles listed in chronological order the entry on the *Cosmologische Briefe,* described as a "work full of genius and of insight."[176] In Delambre's history of astronomy during the eighteenth century, Lambert received but a fleeting mention,[177] irrelevant to his cosmology, although the work consisted of the summaries of the contributions of fifty-six astronomers of whom a dozen had been certainly less entitled than Lambert to such an attention. There was no reference to Lambert in Laplace's speculation about massive "opaque" stars in his *Exposition du système du monde,* a work which though first published in 1796 made its real impact only during the nineteenth century.[178]

It should seem even more revealing that W. Olbers, whose method of calculating cometary orbits was an elaboration of that of Lambert, explicitly rejected Lambert's claim about a teleological arrangement of comets.[179] But at least he referred to Lambert. F. T. Schubert, astronomer of St. Petersburg, presented in his *Populäre Astronomie* a two-page-long account of the general structure of the universe, an account which contained all the Lambertian specifics, except Lambert's name![180] F. G. W. Struve, the foremost figure of astronomy in Russia during the nineteenth century, was far more careful in such matters, but he was too keen an observer not to spot the observational shortcomings of Lambert's cosmology. They formed a sobering contrast to Struve's spirited summary in twenty paragraphs of Lambert's views of the stellar heavens,[181] and especially to Struve's claim that Lambert's cosmological work was "remarkable by its clarity and exposition and by the incisiveness of its views," and that one could not but admire Lambert's "clear and logical deductions, rich also in happy insights especially when one considers the scarcity of solid foundations provided at that time by observations."[182] But, as Struve himself had to admit, Lambert's handling of some observational data was not at all careful.[183] Nor for that matter was Struve's assertion careful that Olbers' cometary researches consolidated Lambert's views on comets.[184]

That further progress in computing the orbit of comets was a support for Lambert's method of estimating their number was, however, soon forthcoming in a systematic manner from a most prominent source. Using two lists, one from 1831 with 137 computed orbits, and another from 1853 with 201 orbits, F. Arago made a statistical projection of their total number in the solar system. For a basis of that projection he took, as he put it, the "much more elevated considerations with whose help Lambert once tried in his *Cosmologische Briefe* to achieve the solution of this curious problem," the object of a twelve-page-long chapter in Arago's posthumously published four-volume *Astronomie populaire*.[185] The chapter, "How many comets are in the solar system?", was, apart from a section in Struve's book, the only thematic discussion of any major cosmological idea of Lambert from the pen of a prominent astronomer during the nineteenth century. Arago's conclusion was that the number of comets increased with the cube of distance and not with its square as Lambert had claimed. Needless to say, Arago, well known for his positivist preferences, passed over in silence the broader cosmological perspectives of Lambert's interest in comets.

Progress in the cataloguing of stars during the first half of the nineteenth century earned for Lambert's ideas on the realm of stars not even such an isolated and short-lived triumph. The eclipse of Lambert was all too plain in a small monograph by J. H. von Mädler, director of Dorpat Observatory, on the central sun which, when published in 1846, created quite a sensation. There Mädler recalled that Argelander's and Bessel's work on star catalogues definitely showed that Sirius could not be a central star and that there remained but small likelihood that the system of stars had a central "ruler." Both theoretically and observationally, Mädler argued, it had become obvious that groups of stars could very well exist without the gravitational influence of a massive central body. Such statements were as many natural pointers to Lambert but he was not once referred to by Mädler.[186] In discussing the future tasks of astronomy, Mädler singled out as a prominent one the verification of a universal center, a project largely depending on the application of photometry. It was in that connection that Lambert finally received mention. Photometry, Mädler remarked in a footnote, was "no longer in the hopeless condition in which it was at the time of Huygens and Lambert."[187]

Lambert was simply referred to, together with Wright and Kant, in the *Kosmos* of Humboldt—who often quoted Struve—as one whose ideas about the spatial distribution of matter were placed by Herschel "on the solid ground of observation and measurement."[188] Clearly, one could not expect Humboldt with his strongly positivist bent of mind to say something reliable about Lambert's speculative cosmology, let alone to create interest in it. At any rate, when the first volume of the *Kosmos* was published in 1845, the principal issue in astronomy and cosmology, the question of island universes, hinged much more on data provided by the giant telescopes of Lord Rosse than on speculative principles.[189]

Lambert could not be seen as a forerunner in any sense of the two main interests for astronomy and cosmology during the second half of the nineteenth century, stellar spectroscopy and the question of the evolution of planetary systems. It was only the rise of interest in the history of astronomy, in Kant's philosophy, and in German scientific past, that directed attention to Lambert in Germany during the closing decades of the century. The only reliable account of his cosmology published around that time failed to reverse the tendency to lump Lambert with the "Kant-Laplace" theory.[190] His obsession with comets, dark regents, and cycloids was invariably overlooked. True, he gave a fairly correct explanation of the Milky Way, and he spoke of many galaxies, but he did not anticipate their enormous varieties as they have appeared through telescopes since the time of Herschel. What might and should have been recalled about Lambert, the cosmologist, was his uncanny realization of the contradictoriness of the infinity of the universe, an insight which he based in part on the impossibility of an actually realized infinite number.[191] Infatuation with the infinity of the universe had, however, been so great, that general credence was given to a dichotomist view, according to which the whole visible universe was confined to within the Milky Way and the still infinite realm of matter beyond it was inaccessible to and irrelevant for science.[192]

Unlike the nineteenth century, the twentieth was no longer under the complete spell of the infinity of the universe. General relativity and its cosmology have no room for Kant's naive canonization of infinite Euclidean space and universe. Whatever the theoretical possibilities of a universe with infinite mass,[193] it cannot have a zero curvature or Euclidean infinity. While among the various models, the one with a curvature corresponding to a universe finite in mass (space) and time seems to receive the best support from observations, several efforts have been made to reinstate infinity. One of these moves consists in assigning to the universe a hierarchical structure. In discussing the optical paradox of an infinite Euclidean universe of stars, the younger Herschel had already noted in 1847 that a specific spacing of stars or galaxies can overcome that paradox.[194] More recently C. V. L. Charlier advocated that specific spacing analogous to a hierarchical ordering, and he indeed referred to Lambert.[195] Very recent efforts to find observational evidence for supergalaxies could have also been motivated in part at least by uneasiness about a finite universe.[196] To be sure, a finite universe exudes metaphysics in a degree which is not palatable to many in a positivist age. It is, however, difficult to see why the singularity implied in a hierarchically ordered universe should seem to be less metaphysical, even if it saved infinity. This first English translation of Lambert's *Cosmologische Briefe* should help recall that a hierarchically ordered universe was for its first proponent a mighty pointer to metaphysics. The full text in English may also be a help for lessening the occurrence of stereotype, faulty accounts of details relating to the history of astronomy, including its most recent phase. With the full text in hand, one may, for instance, get a better grasp of the reliability of comparisons between the "dark regents" of Lambert and the "black holes" of modern relativistic cosmology.

IV   Translating and Annotating the *Cosmologische Briefe*

If the comparisons mentioned above overshoot at times the limits of soundness, it is partly because undue modernity is given to expressions which when used centuries earlier could only reflect the mental world of bygone times. Faithfulness to the historical milieu is all the more important in the case of an author like Lambert who aimed at promoting a German academic style free of foreign words,[197] especially those of Latin and French origin. Such words stand out in Roman script in the text printed in Gothic and are few and far between. Although this could contribute to the ideal of a purely German vocabulary, exactness and clarity were not always helped. While "Erfahrung" lends itself to more specific connotations than "experience", "Observationen" which Lambert used sparingly would have fitted far better in most cases the generally astronomical context. Curiously, Lambert hardly ever used "Beobachtungen." His was a rather indiscriminate use of "fixed stars" and "stars."

Though many examples could be given about Lambert's peculiar and redundant use of words, the problem it creates for the translator should seem small compared with the difficulty posed by Lambert's art of joining words, or his syntax. While he shunned words of Latin origin, all too often he was engulfed in cumbersome, long phrases, which only if used by

Cicero would have retained smoothness and clarity.[198] Lambert was unabashedly fond of long and complicated phrases. They were to carry the touch of pathos which he certainly intended to be a part of his discourse on cosmology. As was already mentioned, he defended his long phrases and he did so in connection with a projected translation of the *Cosmologische Briefe*. Cutting those phrases into short sentences would, he thought, deprive them of inner strength. In this translation a scrupulous effort was made to render Lambert's phrases as faithfully as they are in the original and not according to canons set by dogmatic editors concerning the presentation of scientific topics. Lambert's phrases, which are short only on occasion, are often baroque and stilted. To make matters worse they often carry the marks of hasty composition and of typesetter's errors. When such phrases were improved in the translation, the words added for the sake of clarity have been included in brackets. His manner of using numbers was retained intact. Although a proponent of a hierarchical universe, he never used the word hierarchical.

Clarity and logic were not helped in the *Cosmologische Briefe* by its author's choice of the epistolary form and of two correspondents whose thoughts differ but little. Indeed, with the exception of the sequence of Letters I–II and IX–XIV Lambert seems to correspond with himself, a circumstance which makes his line of thought appear elusive in places. A somewhat similar problem arises from Lambert's intention to keep his discourse on a fairly generic level. As has already been noted, he wrote with only a general view on what had been proposed in the cosmological literature of the previous hundred or so years. Consequently, a certain discretion had to be exercised in offering notes to clarify that broader background of his reasoning.

Lambert's specific statements and references are literary, historical, and technical. The literary ones are very few in number. Since Lambert's cosmology has now mainly a historical interest, notes concerning points of the history of astronomy and cosmology were given special attention. It is here that belong references to notes offered to Darquier's translation by Utenhove who was eager to put Lambert's ideas into the context of the progress made in astronomy during the last four decades of the eighteenth century. The two chief aspects of that progress were Herschel's investigations of the Milky Way and the studies of planetary perturbations by Laplace. Utenhove's frequent references to Laplace's nebular hypothesis are another indication of his alertness to the latest in astronomy, a fact all the more significant as it was not until a long time afterwards that the nebular hypothesis found a proper echo in France in particular. What adds further significance to Utenhove's notes are his repeated allusions to the original edition of Kant's *Allgemeine Naturgeschichte*. He seems to have been the first to write in some detail in French on Kant, the cosmologist, and also the last for almost three more generations.[199] Notes of technical nature are relatively few because Lambert's line of thought is never too technical. It can also be seen without technical notes on Halley's Table that no cosmologist has drawn as many inferences with as great a confidence on the basis of so few data as Lambert did.

The main presuppositions are clearly and even repetitiously stated in the *Cosmologische Briefe*. They are Lambert's cosmological principles,[200] unmistakably philosophical and a priori in character. He stated unabashedly that as the product of an all-wise Creator, the universe had to be the most perfect of all possible worlds, that such a perfection entailed freedom from cosmic cataclysms, and that it meant the presence of life everywhere in the cosmos. This in turn implied the existence of as many abodes for intelligent life as possible. Such a consequence could only be implemented by a very large number of comets around each star and by an immensely large though strictly finite number of stars. This finiteness of the universe was for Lambert a consequence of its purposeful arrangement, which had to reflect the same ordered, or hierarchical, organization throughout. Some of these presuppositions, especially the belief in the existence of denizens everywhere in the universe, was widely shared even by Lambert's younger contemporaries. Here Lambert offered nothing that caused surprise which is, of course, no indication that the belief itself was distinctly rational even when taken in the context of the times. As to teleology, there was already developing a deep cleavage between those who still accepted it and those who had already turned their back on it. Lambert provoked as little comment with his explanation of the Milky Way as did Wright and Kant. Of all major conceptual breakthroughs in cosmology none entered general consciousness more unobtrusively than was the case with that explanation. Such is possibly the reason why Herschel, a full generation after Wright, Kant and Lambert, created more sensation with his count of nebulae than with his work on the shape and structure of the Milky Way.[201] While much progress was made in the observation of comets during the closing decades of the eighteenth century, it was never evaluated, apart from Utenhove's comments, in the perspective of Lambert's philosophical reasoning on behalf of their great number and specific arrangement. His advocacy of the finiteness of the universe could strike no responsive chord at a time when even the optical paradox of the infinite universe was largely ignored or talked away in a distinctly facile manner.

As to the concrete scientific supports of his cosmology, Lambert himself stated on many occasions their incompleteness. He wrote with a view on future developments in astronomy which, however, failed to justify his hopeful vistas about the truth of much of what he had submitted about the arrangement of the world-edifice. It certainly became clear that comets, though very large in number, were anything but lasting celestial bodies, and certainly no place for astronomers, however superior in mind and body. Although thousands of new comets had been observed since Lambert's time and Halley's one-page-long table of comets with computed orbit had since swollen into a thick volume, cataloguing of comets did not become a key to cosmology. The nebula in Orion did not turn out to be the indirectly illuminated surface of a "dark regent," and the "black holes" of modern relativistic cosmology did not prove themselves to be "central rulers" of vast arrays of stars, let alone of galaxies, all hierarchically ordered. It is in fact very doubtful that a hierarchical organization, which is observed on the level of satellites, planets, and stars, extends to the level of galaxies. Last but not least, the direct study of the moon's surface, together with data provided by space probes sent to Venus and Mars, and around the more distant planets, may suggest that the earth could very well be a much more unique abode than present-day claims about extraterrestrial life would have us believe. Such claims are at times as heedless of indications to the contrary as was the manner in which the principle of plenitude was upheld for comets in Lambert's time.[202]

One may therefore be tempted to say that Lambert merely provided an example of how far one could go without the support of observational breakthroughs and how dangerously far one could advance. But to his undying credit he dared to advance into the unknown by conjuring up a vast, systematic image about the universe. Undoubtedly he speculated too much on the basis of too little, but such is a risk which has to be taken time and again if the gathering of data is ever to become a meaningful science. The real significance of individual data never emerges until they prove or disprove a broad theory, and nowhere must a theory be broader than in cosmology, the most inclusive field of scientific investigations. Indeed, no individual fact can reveal more of its scientific meaning than through the effort to give it a basic role in a cosmological theory. Such an effort is a continual exposure to success as well as to failure, with the latter being usually predominant. It is this kind of mixture which makes the history of science, and this feature of it is nowhere more pronounced than in its primary texts or classics. One of them is the *Cosmological Letters on the Arrangement of the World-Edifice.*

Cosmologische Briefe

über die

Einrichtung

des

Weltbaues

Ausgefertigt von

J. H. Lambert.

Augspurg
Bey Eberhard Kletts Wittib
1761.

COSMOLOGICAL LETTERS

ON THE

ARRANGEMENT

OF THE

WORLD-EDIFICE

Drawn Up by

J. H. LAMBERT

AUGSPURG

at Eberhard Klett's Widow

1761

PREFACE

I was prompted to publish this collection of cosmological letters on the arrangement of the world-edifice by two considerations which, I think, should be pointed out here because, according to the readers' various ways of thinking, these letters must of necessity make different impressions on them depending on whether the propositions set forth therein, and especially their proofs, are more or less clarified.

[IV] In that section of the *Photometrie*[1] which deals with the light and distance of fixed stars I have incidentally remarked, I believe, that one can represent to oneself their distribution in cosmic space in a most reasonable manner.[2] I have summarized the most important considerations on this point and contented myself with presenting them in a few statements and without proofs, although one could have demanded these from me all the more because the notion which I submitted was rather unexpected and could be considered a consequence of several conclusions.

While more recent astronomers are very anxious to put in order the sphere of our solar system and to assign each comet its orbit, and to determine in advance its eventual return, hardly any attempt has been made to find out something probable on the arrangement and position of the fixed stars.[3] What has been done in this respect was that one tried to figure out how distant the nearest of them is from the sun.[4] Undoubtedly, one can hardly infer from this the arrangement which they have with respect to one another. The measurements in this connection do not reach far enough and one must therefore necessarily employ general considerations [V] which, though they do not possess geometrical precision, can nevertheless be brought to a rather high degree of probability.

These considerations form the contents of the last letters of this collection. They deal with the arrangement of the whole world-edifice. The earlier letters are devoted to its individual parts and especially to determining the arrangement of globes in our own solar system, which in part is already known to us and will, through future observations, be even better known. The purpose which animated me there bears on the fact that I have shown that the comets are far from being so fearsome as some would for some time have had them to be; that our solar system is not so empty as one might have believed due to lack of sufficient observations; and, finally, that while in respect to dignity comets do not outrank planets, they are not in the least inferior to them.

I know that with all this I propose many new thoughts, and with great boldness at that, as if this boldness should make up for what is lacking in the precision of the proof. Such was not, however, my intention. I admit that most of what [VI] I say here has only a certain measure of probability. This I tried to stretch with careful reflections as far as possible. I approached each issue from every side to find new grounds of support, and I have sought to give the proofs all breadth and to bring them as close to certainty as the matter permitted. I investigated what belonged to full certainty and put the reasons together in the balance to see what was still lacking to their full weight. I considered them with an eye on the readers and pondered how much would be granted to me by each of them according to his own way of thinking. This is all I could do with propositions which were but probable. I did my best to gather for the future a supply of probable proofs which I would gladly see to be greater. And now my second purpose.

For many years I have been busy with the task of abstracting the techniques and rules as they come along both from my own findings and from those of others, so that I would not easily miss any which I have not looked for, and thereby to make for myself [VII] a collection of them which in the future I would present as notes and addenda to a theory of knowledge and discovery.[5] Here also belongs a part which is devoted to arguments, inasmuch as these are opposed to demonstrations having full force and might possibly be looked upon as insufficient proofs. It is unnecessary to set forth here their various kinds and how far each of them would reach. At any rate, the examples in the collection mentioned above will be all the more useful to me.

It should be clear that the present letters give a good supply of those examples. It has since long been desired, concerning the philosophical analysis of the probable, that rules and examples, insofar as they concern probable things, should be worked out. I do not find myself in the situation of giving such rules and examples which might be used in ordinary life. But since ordinary life does not exhaust the realm of probable reasons and since many of these are still unknown even in the sciences where one should have sheer certainty, these letters should in a sense be a breeding place for those reasons which deserve [VIII] to be scrutinized all the more sharply, because teleology should demonstrate to us in the study of nature not only the generality of nature's laws but should also be very useful for finding them.

This latter usefulness was already pointed out by Mr. von Leibniz who sought to give an example of it with the proof of the law of refraction of light rays.[6] Mr. von Maupertius also endeavored to demonstrate all laws of motion from teleology.[7] In general, it is indisputable that there should be more than one maximum and one minimum in the world. It is regrettable that the study of teleology is still a matter of sheer faith, so that everybody reserves to himself the right to have his success in an arbitrary fashion;[8] and it is also a pity that concern about exceptions from the generality of teleological tenets is carried to such an extreme that one would rather have those tenets confirmed through experience and not trust them before seeing them at work.[9] This difficulty affects most of the proofs of which I avail myself in these letters, so that, unlike the geometrical proofs, they do not dispense of approval; rather it remains to the reader [to decide] whether he finds them worthy of approval. Many presuppositions of which I make use in these proofs are derived [IX] from the purposes of creation, and are consequently teleological. One may call in doubt their generality in two ways. One may ask whether they are not necessarily subject to strong limitations, and even when such is not the case, whether there are no individual cases where other purposes would pose exception to them. The body of doctrines about the purposes of creation is far from being so complete that one could compare each single purpose with the rest, firmly establish their subordination, determine their limitations and exceptions, and evaluate each case on that basis. As long as this is not so, each reader will claim for himself unlimited right to endorse and to reject as much of teleological proofs, as pleases him. One shall find in the Sixth and Eighth Letters considerations which have to do with this. In the former there will be submitted a complete and well proven teleology; in the latter there will be an investigation of the right which each reader may claim to himself according to his specific way of thinking to doubt the arrangement of the world-edifice here described.

A principal difficulty which is inherent in the teleological proofs is the finiteness of the world. Great [X] and immense as the world may appear, it is restricted to finite numbers,[10] and thus half of the infinity of divine purposes is eliminated. Their sum can be considered infinite only with respect to time, because they extend themselves over eternity and therefore always approach the infinite though without ever reaching it. With respect to space the situation is different. Here the whole has its limits which curtails the generality of teleological tenets, so that one must always add the condition: *The world-edifice reaches so far.* Therefore one cannot submit them as unlimited, and it is easy to note that this provides ground for doubting the appropriateness of the application. Thus, for instance, if one wants to consider the habitability of the world as a principal aim of creation and to derive conclusions from it, one cannot assume this without the limitation noted above, namely, that the habitable places must be limited to a finite number. One should always present that aim only in the form: *as far as the world-edifice reaches, it is inhabited.* It is incontestable that the proof of this tenet is more difficult than if one were not to assume either exceptions or limitations concerning that purpose.

[XI] I cannot say whether all that caution is employed in all teleological proofs occurring in these letters, because I have simply submitted most of the premises as being generally accepted, since this is usually done in probable demonstrations to secure their full strength. One may therefore look at them as *loci communes,*[11] the like of which are often made use of in every-day life. I let you have this form of proof without any hesitation, because the whole form of proofs is structured accordingly. I postpone the closer analysis of each kind of these proofs to my promised comments on the theory of knowledge without submitting to scrutiny any of them in these letters.

What I would say here about them is that I have sought out such proofs *as have already often occurred in similar cases.* Among them one may reckon the doctrine of antipodes,[12] the doctrine of the earth's rotation, the thoughts of Seneca about comets and his predictions about them,[13] the doctrine concerning the inhabitants of planets, and several similar ones, which experience either readily confirmed and made apodictically certain, [or] in part raised to much higher degree [XII] of probability. One should think that proofs of this kind, of which each considered in itself is too weak but all of which taken together have an illuminating strength, constitute a special kind of certainty which, though different from the geometrical, is not at all less certain, and the difference of this certainty rests perhaps merely on the fact that such proofs are not yet brought into their proper form, that the strength of each cannot yet be calculated, and therefore it cannot yet be seen whether their sum makes a whole which implies complete certainty. This imprecision is enough to let a proof, which is more than satisfactory, appear in most cases either to be unsatisfactory, or at least doubtful.

To give the teleological grounds of which I have made use a greater authority, I have not omitted to derive from them such consequences as well which experience has already clarified. One can list here as an unexpected example what will be said about the condition, position, and number of planets and cometary orbits in the Sixth and Seventh Letters where I infer in a necessary manner from the greatest possible habitability and adroit separation of the globes [XIII] in our solar system, and therefore from teleological ground, that there are many more comets than planets, and moreover that the planets must lie in the same plane, and the outer planets should stand farther from one another, exactly as experience teaches us.

Still not all grounds which I use are merely teleological. The law of gravity which I extend through the whole world gives me grounds that lead to conclusions in a much more necessary manner. The separation of fixed stars from one another will be derived from that law in the Twelfth Letter, and the mutual avoidance of planets and comets in the Third Letter. In consideration of the latter I only have to recall that the proof, which arises in connection with the question whether a comet can turn into a moon, will for brevity's sake be so offered as if the earth were to remain fixed at a point in its orbit, though this cannot be assumed rigorously. The moon describes in its true course a cycloidal line in which it moves not much faster than the earth. At new moon its motion is as much slower.[14] On the contrary, the motion of a comet, when [XIV] as far from the sun as the earth, is speedier almost by a half, and therefore in its return toward the earth this speed should of necessity be greater, so that it is impossible that the comet could be captured and hence accompany the earth forever as does the moon. On the other hand the moon is in a stable condition into which no comet can come, because the means of bringing a comet into the cycloidal path of the moon simply belongs to the realm of the impossible and cannot be thought of without a true miracle. This had to be remarked here to indicate how the proof had to be constructed.

What I further deduce from customary teleological and other grounds, I have tried to extend so far that most of it could be clarified sooner or later through experience and careful observations. Here also belongs what I say about the number of comets which I made, largely through assumptions, to appear as considerable as this could be done to show that there are good reasons for rescuing these celestial bodies from the former low esteem into which Aristotle and his followers[15] placed them and for considering them once and for all in their true dignity, [XV] inasmuch as I show that they are much more necessary and useful for the habitability of the solar system than are the planets, and have in addition a greater variety in diversity, and therefore contribute the most not only to the completeness but also to the perfection of the solar system, this parcel of the whole. It is enough if one agrees with what I say about their number only to the extent that this ought to be very considerable and that the light and warmth of the sun is indeed utilized to the utmost. The true determination of these matters will be reserved to future times when Halley's Table,[16] which I consider useful to add in order to shed more light on the frequent remarks on them, will be made into a complete register of those comets that come into view from our earth. From such comets one may then draw an inference about those which remain farther from the sun whether they ever become visible to us, whose number I set at least 40 times greater even if no perihelion is more distant from the sun than Saturn, the outermost of planets visible to us.

In the second half of these letters I have in particular tried to give a considerable magnitude [XVI] to the world-edifice, and here it especially seems to be no small boldness to overcome the lack of stringent proofs. I can readily leave to the reader how much of it he will grant, especially when one does not make the effort to consider the proposed observations and calculations, or when such an effort should yield no fruit. At any rate, I can find no reason to reject completely these bold thoughts, much as they might become a mere matter of faith, and I group without any hesitation into a special system the fixed stars visible to us, which taken together constitute the Milky Way. This presents to me a whole system which presumably belongs to still several others. Inasmuch as I can present the matter only as probable and leave the more doubtful details undetermined, there still remains at any rate a wide field for speculation which each reader, who does not consider the undertaking unworthy, can further pursue at pleasure.

To these I add in the last letters such suggestions as at the outset I myself did not entertain, because I thought to stop with the Fourteenth Letter, and having had the work completed up to that point I laid it aside for a while. About the truly observed [XVII] displacement in the position of stars with respect to one another nothing was known to me. I have derived it only from general reasons which I submit in the Tenth and following Letters. One shall find that these reasons are almost exactly the same from which one had already demonstrated several propositions in astronomy, or brought them at least to an illuminating degree of probability. One forwarded them in a general fashion and I could find nothing through which their generality might be limited. It pleased me all the more to find that the displacement of fixed stars deduced from those principles was in a good many cases[17] confirmed through observations, and Herr Prof. Mayer, who earned much honor with this discovery as with all others for which we owe him thanks, himself does not doubt the universality and validity of the complete induction.[18]

Since verification came sooner than I myself expected, it led me to develop in greater detail my conclusions, which until then had naturally followed one another, to see what else they could still disclose when I considered them in a similar generality and carried the analogy even further.

[XVIII] Here I shall leave it to the sheer pleasure of the reader to judge whether I have constructed for myself a *deductio ad absurdum,*[19] if the proposed observations, for which I had then neither time nor opportunity, were not to turn out according to my expectations.

What I still think prior to the indicated proof about the whole matter, I have brought together in a brief paragraph in the Twentieth Letter and leave altogether to the reader what he will grant or reject of it, because the matter prior to observation is simply a topic of faith, and the unbelievable in it pertains not to the impossibility but merely to the unexpected and the unusual. Had the observations already been presented, I would have either left aside [the topic] or proved it better. But I find it not at all unusual if, with a view to a future demonstration, I risk an opinion and raise such [logically] connected questions, which observation may answer. They are of such a kind that whatever the answer happens to be, it will prompt further investigations.

Propositions that are made only in a probable fashion differ from others which are demonstrated [XIX] in all rigor, especially since in connection with the latter one rarely deems it necessary to think about answering objections which mostly arise through misunderstanding, whereas with the former the rebuttal and answering of objections implies much more, because often the whole proof of a probable proposition rests on one's seeking to show that nothing, or at least nothing of importance, can be objected to it. A correctly proven proposition is necessarily and properly bound together with other truths; with probable propositions the connection is either not wholly necessary, or is incomplete, or at least reveals nothing that would overthrow it. Should they be refuted more rigorously, one has to determine what is still lacking for a complete proof if one reduces this to the most obvious and available data, and then [one has to see] whether it is still insufficient, or not appealing enough, and how far one could accept it without stronger proofs.

Should the propositions that are considered from all sides in the latter half of these letters be placed in such a balance, then it readily obtains that they are of different weights. In particular [XX] the proposition about the motion of fixed stars around a center is also demonstrated completely because it rests on experience, on the first laws of motion, and on the preservation of the world-edifice. Experience shows that there is really some motion. The maintenance of the world-edifice eliminates the linear motion, and the fundamental laws of mechanics in addition to the laws of gravitation make that motion completely central.

The other main proposition, namely, that the Milky Way is divided into smaller systems of fixed stars and that these are separated from one another by considerable spaces, lends itself to a less solid proof. What is lacking here is that we see these systems not from all sides but only from a specific angle, in which the Milky Way clearly shows itself separated from other parts of the heaven and divided here and there into smaller parts. At any rate, such a situation should still be displayed even if we could see the Milky Way from other directions. But since this is not a possibility, the incompleteness of the proof should be supplemented from other considerations as far as this can be done.

I can submit the third main proposition in no other way than in the form of a problem, and I consider its solution [XXI] as yet very incomplete. The question in particular is whether cosmological or mechanical grounds allow that a system of fixed stars should only move around its center of gravity, or whether there should be in these centers, as is the case with smaller systems, bodies of proportionate mass. The mechanical grounds rest on the laws of gravity, the cosmological grounds rest, however, on the maintenance and simple order of the whole. Should one answer this question affirmatively, one cannot very well avoid the [existence of an] enormous [body] surpassing all power of imagination, and one has to admit this as soon as the question is to be answered affirmatively. In that case one has a close though not complete right to press forward to an autopsy.[20] If, however, one can answer the question not in a necessarily affirmative manner, the problem remains inevitably similar, because the opposite cannot be made in and for itself necessary as the smaller systems embody examples to the contrary. And the power of imagination, which finds in it something too enormous [and which] has the proper right to set its own confines as the limit of the mark of truth, has failed many times before and in far more important issues. It seems that it falls behind both in big and small matters.

[XXII] What may further inconvenience probable propositions, if one sets them forth in such a manner, is that both the foundations on which one builds and whatever one connects with them, are mere matters of faith. This makes them all the more hypothetical, because one has to assume that the utilized foundations will be proven as time goes on or at least will be accepted as trustworthy without much objection. It is easy to show that the foregoing third main proposition[21] is of such kind. It can be scrutinized only conditionally, because it presupposes the first two main propositions as wholly correctly demonstrated. Should, for instance, the arrangement of the fixed stars into systems not be granted, then the question whether these systems have in their center a body of proportionate mass would not rise at all, and much less would the question pose itself whether these bodies, which together form an even greater system, should have in their center an even greater body.

In much the same way that these questions seem or may be posed can the reader, who wants to investigate more carefully the whole subject matter of these letters, marvel at the boldness [XXIII] that made me set forth these questions in an order as if everything had already been settled in an incontrovertible manner and assumed accordingly. I might have stopped and stayed with the first two main propositions, and perhaps many would judge that I should have done so. However, I sincerely believed it to be my duty toward the readers to show them what I myself foresaw and to set forth the questions in a sequence in which they logically follow one another. Undoubtedly, the last ones in the chain of these questions are the least proven, but they can nevertheless help make more useful the investigation of the first ones. Moreover, the decrease of the measure of probability and credibility in the latter part is gradual, and for that reason also I found no cause to cut the chain somewhere abruptly. I have therefore set them forth in full. They can in no other manner but in proper sequence be investigated and brought into the clear, and before answering affirmatively the first ones it seems superfluous to argue over the last ones. Should all be answered affirmatively, we obtain a more complete doctrine about the arrangement of the world-edifice, and with this presupposition in its truth I have tried to expound this doctrine in the last three letters in its fullest harmony. Should, [XXIV] however, these questions not be answered affirmatively, the whole doctrine will more or less fall apart, and we shall be left with the pleasure that it is even more harmonious to be forced to find out more laboriously what we do not know.

With respect to the whole world-edifice we seem to be much in the position in which Pythagoras, Philolaus, Aristarchus, Nicetas,[22] and other Greek sages were in respect to our solar system. They risked speculations, which they did not completely elaborate partly because of convenience, partly because of lack of satisfactory observations. Perhaps even the so-called Ptolemaic system was not so sufficiently known that one could derive from it proper reasons for overthrowing it. Alphonse[23] seemed to have more reason to be impatient with it without having the good fortune to turn around the whole matter. This was reserved to the immortal Copernicus, and Kepler and Newton had to complete the process. We are still waiting for the Copernicuses, Keplers, and Newtons of the whole world-edifice, and we can formulate the prophecy similar to the one which Seneca uttered in connection with [XXV] the comets.[24] It remains to be seen whether fulfillment in this case shall come sooner.

I could only wish that I turned the considerations in these letters into a second part of Fontenelle's dialogues on the more than one world,[25] provided I could have presented them with that precious liveliness, and [assuming that] the inspiring notions had come to me in a similarly fluent and rich fashion. Each reader, when he sets aside the vortices used so often and so readily by Fontenelle,[26] can easily see to what extent what I say about the world-edifice may in principle serve as a continuation of Fontenelle's ideas. How much more so, when, instead of finding here only one world, one grants a legion [of worlds] which are around our sun alone, when each fixed star has so many, when the Milky Way has an uncounted army not only of fixed stars but of entire systems, when finally the whole Milky Way belongs still to innumerable other Milky Ways or to systems of similar kind!

[XXVI] The considerations of Fontenelle will naturally be brought to a certain completion when one does not go beyond considering this army of Milky Ways as a part of still greater systems, which are still puny compared with the infinite. Farther I will not stretch the power of imagination which for me is too limited anyhow, since I had to gauge it with the one which provided Mr. Fontenelle with such lively scenes.[27] I remain within narrower bounds and for this reason I prefer the form of letters to conversations, because I had to set forth many considerations in a mutually connected manner and because in such demonstrations, which ought to be somewhat more rigorous, dry details could not be avoided.

53

I may just as well leave it to the critic to compare the character of the two friends to whom I gave the pen in these letters, or to investigate how far I myself am removed from the whole matter. In matters in which the material is the chief issue one will not be too preoccupied with this.

[XXVII] Of these friends, the one who replies always remains himself, while the other must change somewhat because he did learn and became more accustomed to thoughts connected with one another. I could not make the two friends be of very different frames of mind, and I have already given the reason why objections, which might have occurred to some readers, had been left aside, for I rather sought to explore the reasonings from every side and make them variegated. From this it will be clear why it is not important for the student to press his objections too far.

The letter-form demanded expressions of praise and of courtesy which I would have otherwise avoided had I given these as my considerations. One may look at them as resting places that interrupt more tiresome undertakings. In particular I have chosen those expressions which illustrate either the duties or characteristics of true friends and represent their way of thinking, or [such expressions] which in discourses of logic [XXVIII] touch on fine points of understanding, [or] which, although applied here as praises, will do honor to any reader who takes them, and which make him as a friend lovable and as a searcher of truth worthy of respect.

COSMOLOGICAL LETTERS
ON THE
ARRANGEMENT OF THE WORLD-EDIFICE

FIRST LETTER

Would you believe, Sir, that after reading the material you have sent me I feel more uneasy than before? I hoped to satisfy in full my curiosity on the orbit of planets and comets, and I congratulated myself in advance about the good fortune that I soon shall know in a reasoned way how it happens that comets return in appointed times; and this [2] pleased me all the more because during the past year we could celebrate the first return of such an extraordinary star.[1] I should indeed be very indebted to you. I perused these lessons with the greatest interest and nothing is now easier for me than to fill with oval lines the vast spaces around the sun, all the spaces between the planets and even those far beyond Saturn, and to attach to each of those ovals a comet, and with satellites if you wish. To each fixed star I gave a similar crowd of such bodies which must receive from it their light and heat, and on each of these I placed innumerable inhabitants of all possible kind and form. I have thereby stretched the imagination as far as the world-edifice reaches, and it is now no problem for me to take the distance of our sun from a fiftieth magnitude star as a yardstick, and, by laying it off a million times, to set it up as a measure against the limits of the system of those stars which we see with telescopes and of those which are still beyond. I no longer use an anvil which takes ten days to fall from heaven to earth.[2] The space through which it falls is now to me a mere point, and I compare its velocity with the crawling of a worm or with the creeping of a snail. Should I compare time and space, the flash, the lightning which spreads across the sky in an instant, is still too slow. Only the light and the distance it covers do for me [3] as a measure. It comes in eight minutes from the sun to the earth and leaves behind a path, which one has to measure with earth-radii of which each is equal to 860 [German] miles and of which one needs at least 20,000 to determine the distance of the sun. The light covers that route in eight minutes, and such is now my yardstick with which I locate the farthest fixed stars. I give the light centuries of time to come to us from those stars, and I posit that there are fixed stars from which the light has not yet arrived during the past 6000 years[3] and consequently they shall be seen only by our descendants. The night should in my view become always brighter and each night I rejoice at the newly arriving light from other stars. For all that I see that such distances still have their shortcomings and that they are perhaps still far removed from the boundaries of the world-edifice.

I may be taken in unwittingly by the charm of astronomy and I may be talking to you, Sir, of things that are well known to you. Accept these always as a proof that I have followed your advice and I tell you again that it gave me much pleasure. But don't you also tell me that it is not possible to reach certainty in another way than through new doubts and new uneasiness, or do doubts merely change their nature in that they are simpler at the start [4] and become involved once they have been solved? At the start I have only asked about the structure the world-edifice may have, but now I know that the question remains to be asked as to what shall become of it as time goes on, and what shall become of the philosophers who have so pleasingly figured out all this. Is it not that comets are no longer fearful through their significance but through their effects?[4] I know sufficiently only the forces of gravity to which celestial bodies are subject with respect to one another. Jupiter can disturb Saturn and its satellites, and even the moon can sway the earth. It makes the sea rise and fall. What would happen to us were a great comet to come close to the earth and raise the sea above the dry land or to carry along the earth altogether?

Tell me now, Sir, whether one should simply hold as probable all that was proposed by the philosophers in that respect, or do they unwittingly mix with truths such possibilities which should fill one with fear, were they really true? Actually, after I have considered all these horrendous possibilities, I was noticeably slow in spreading comets freely through cosmic spaces and I have dropped all those that could bring about a catastrophe as time goes on. Still with all that I do not get too far because I must take comets as they really are, and when they themselves do not occupy [5] peaceful orbits we are always to worry about warfare in the firmament.

What do you think about this? Can one find apart from sheer possibility also something probable in such wild events? Might Jupiter, Saturn, and the earth have indeed grabbed their satellites in such a warlike manner that they had attracted comets and carried them along, though these could have continued quietly in their elliptical orbits? How do the inhabitants of a comet fare in such a new position? And had things to pass in connection with the creation of the earth and with the flood in such a way as Whiston,[5] and Burnet,[6] and others[7] describe it? I tell you right away that all this strikes me very much like a novel, a quality which one should not seriously seek in philosophers. The writer of *Noah*[8] could very well avail himself of that quality, and poets can be allowed the freedom to give free rein to their imagination. So I read them with pleasure and I easily place in one class the comet [of Whiston], the flying ship, and several similar inventions. A poet can rest content with the possible, and in his world he can arrange everything in whatever way things appear most beautiful to him and most wonderful. However, philosophers must look for more than sheer possibilities, and I must confess to you that I did not look for so many dreams in connection with so many beautiful truths. For I cannot call them anything else as long as they are not made more probable. I leave them readily to poets, [6] and could I prove their improbability I would do all to keep philosophers away from them. But they present all these matters to us in such a way that I find neither probability nor improbability in them; they would not even vouch that today there would come no such an evil comet that would rob us of our peaceful moon, or would give the earth itself a push and leave it in ruins, or at least would leave behind a considerable part of its tail in our atmosphere. I am now almost ready to approve the Chinese, at whom I have often laughed, who set up each night watchmen against the sky just as we do on the fields, and who are watching from their observatories to see if there is a sign of something hostile in the sky.[9] How unconcerned was the good Ptolemy with his resting earth, and how quiet his followers remained until Copernicus came and began to lead the earth around the sun! But now the situation is even more complicated, and Copernicus might not even be too proud of his triumph should he know that we must now be concerned that a comet might come and drag along the earth beyond the fixed stars, or at least would drown, crush, choke, burn all of us and, what is even worse than such calamities, would make us fearful of philosophers. I soon wished that comets had again their old significance and would presage wars and all calamities, which hit us anyhow and are less universal than such effects that [7] threaten not only single countries but the whole earth and come in a much more unexpected manner than do all wars. Moreover, our earth is one of the smaller planets and can all the more easily be carried away. And who knows whether already planets are missing which have departed from the vast space between Mars and Jupiter?[10] Does it then hold of celestial bodies as well as of the earth that the stronger chafe the weaker, and are Jupiter and Saturn destined to plunder forever?

Now you see, Sir, what I am about. You must need know all these fearful threats, because you have indicated to me the books where I have found them; but for all that you seem to be very undisturbed. Is your peace only a certain courage with which you await unperturbed the onslaught of the heavens, or do you laugh at such things as if they were the gimmick of imagination? I am ready to follow you in both [directions] as long as I know what your position is. I appreciate your calm which you have put to the test of all misfortunes, and I take it as an ideal for myself. Tell me only whither I shall follow you, and I will do so with the zeal with which I would persevere beyond the grave, Sir, etc.

[8]
SECOND LETTER

Your letter, Sir, which I perused with pleasure, shows me that in a brief time you have made yourself a clear picture about the world-edifice and about the reliability and unreliability of our philosophers, and in all appearance you have found more than you initially wanted to know. But you should not be surprised that new truths give rise to new questions and doubts. This is the common road along which we come from one truth to another, and it is only regrettable that all this takes somewhat longer than we wished. Do not however think that we shall be stuck with these doubts, and we have the hope that we shall clarify them one after another, though there shall also arise new questions the answers to which are reserved to posterity. Find your pleasure always in what we know for certain and leave future possibilities to be resolved by posterity if at the moment they cannot be tackled.

Still perhaps the speculations of philosophers appear so terrifying to you, Sir, only because they are novel to you, and you will perhaps imperceptibly become used to discourses about such possibilities which surely have not yet interrupted the sleep of anyone. Let us, however, also prepare for the worse. Should the earth face in a brief time such a calamity, what [9] would you resort to? Would you not thank the astronomers and would you not consider them to be prophets enjoined to foretell us such eventualities so that we may cope with them? How, when the earth is to turn into the satellite of a comet and is to be carried by it beyond the sphere of Saturn, or when the earth's surface is to be overflown with water, would you not in the former case prepare for more than a Siberian freeze, or look for a ship in the latter case?

Still I always assume the better and I do not consider these possibilities to be such as to make me think: *hic Deus nihil fecit* [God had no part in that].[1] The preservation of entire celestial bodies appears to me, to say the least, more important than the preservation of such creatures that propagate their species and are reborn year after year. In their case the elder can serve for the growth of the younger, but that worlds should arise from worlds, or that from the ruins of old worlds new ones should again be put together, this would demand much more, as their lifetime is measured in myriads of centuries. These relations you find with things whose birth and demise take place under our eyes. Their lifetime is proportional to their size, and the extent of time which it takes for tulips to bloom and for cedars to grow, as well as the time which is spanned by the life of an insect or by that of a man, have hardly any proportion to one another. These are creatures that propagate their species, but their habitat undergoes [10] only such changes which renew it daily or yearly, and greater changes take several centuries just as a few hours suffice for worms.

Meanwhile, I cannot throw aside the possibilities which philosophers have come up with, more perhaps for their own entertainment than seriously, and I am always inclined to think up more such possibilities of which I could write you more have you not apparently had your fill of them. What one could in all cases most easily grant are such changes through which celestial bodies are somewhat displaced in their course, and these you are already familiar with from the examples which you list. These smaller displacements are, of course, the consequences of the mutual gravitation of planets and comets. The question is whether one should simply consider them small exceptions from general laws, or whether they are in fact also the means by which the course of these bodies undergoes further changes and becomes more stable.

What would you think, Sir, should one be able to prove that these orbital changes are due to careful planning and that all planets and comets have precisely such size, mass, position, direction and speed that they, regardless of the constant mutual attraction, always avoid one another in the [11] most skillful manner? Would it not be possible that a comet, which once passed very close to Jupiter, would be derailed in its course in such a way that while it had previously passed by the sun from the right, would now run around it from the left? The greater were such a change, so much more important would seem to me the causes that brought it about. The cosmological doctrines about the perfection of the world and the teleological principles which we posit from experience about the purposes of natural bodies, are to you, Sir, so well known that I do not even need to ask whether you assume in connection with the setting of the course of celestial bodies just as wise a planning on the part of the Creator as we find, for instance, in all and even in the smallest parts of the human body. To be sure, we cannot specify purpose in [connection with] so great bodies as easily as in the smaller ones where we can survey the sequence of transformations. We see in the heavens at most only the exceptions and it will take aeons until the whole series of transformations will unfold and all its details will lend themselves to mutual comparison. Only then will it be evident what is the meaning of the sum of these smaller changes and how the preceding circumstances fit the succeeding ones.

Am I, Sir, correct in believing that in all this you see only the fearful aspect of the [12] working of comets and that therefore you resent very much their discoverers? I do not think that the philosophers have presented these consequences so fearfully for any other reason than to give something for exercising our power of imagination. You know what premonitions are and [you also know] that impending disorders which worry us pass close by us all the time. The philosophers wanted us to realize that comets appeared in the sky not only to be stared at but that they can also do something. All this was rich material to excite the power of imagination, and it is well known that we necessarily take greater interest when misfortune is presented in its highest degree. Should you now tell someone that a comet can make the year longer or shorter, and make winter from summer, and pull the waters from the ocean over all mountains, and berob us of our moon, or [that a comet can] effect that we henceforth have all year long only one full moon and one new moon, or conversely, that we have every fortnight an eclipse and whatever more there is of similar changes, then you would tell him only mere possibilities which are, however, noteworthy enough to draw attention and exercise the imagination.

Should one, however, ask whether something of this kind would ever come about, then no philosopher will vouch for it, because we do not yet know all comets, nor their orbits, nor their conjunctions. [13] You may here easily think that we shall similarly remain for a long time in such uncertainty if there is no other ground out of which anything more reliable might be determined. I have already outlined to you the general cosmological reasons. Help me now to investigate to what extent they permit application here. In my opinion more special reasons are required in this case, and especially such reasons that let themselves be derived from the law of gravity, by which all celestial bodies are ruled, and from the characteristics of their orbits. The cosmological grounds will only be helpful to infer something about the arrangement of world-systems in general, but the smaller evils, with which the philosophers alarm us, depend in a closer manner on the law of gravity because they in fact can be derived from it alone.

I shall start by posing you two questions which, after reading the essays I have transmitted to you, can easily be solved. Investigate once whether it is possible that two comets, or a comet and a planet, collide with one another, or, once they have come very close to one another, they should rather move both around the sun and around a common center. Here you can easily determine the conditions of both bodies and find how far, for example, a comet should come within the earth's sphere of attraction [14] so as to be captured by it, and where should the comet pass through so as to describe an ellipse of a given shape around it, just as we see this in connection with the moon. You will easily grasp that this specification depends on the velocity which the comet has in its original orbit as it approaches the earth. This velocity must have a specific relation to the closest approach of the comet to the earth, and when the comet does not take precisely that path, the earth will achieve nothing more than to alter more or less the path of the comet. Notice then how artful should be the circumstances for such an eventuality to occur, and whether in such a case one should not presuppose a purpose rather than mere chance.

The other question is precisely what you have already posed to me [namely], whether it is conceivable that Jupiter and Saturn have in such a way acquired their satellites. The solution of the previous question may assist you here as well, and the circumstances which we find with the satellites can render the answer easier. Now all satellites move as do the planets from west to east. Assume then that once they were comets, and there will be only two possibilities for their being captured in the planets' net. Either it can happen that the comet is descending from its aphelion [15] toward the sun, or is moving upwards farther away from the sun. In the first case the comet has to cross the orbit of the planet on the east, in the second case on the west. How great is the probability that the opposite should never occur? For this is what the whole matter is about if one is to posit such an event. There are ten satellites in all,[2] and of those at least five should revolve around the planets from east to west because both movements should be considered equally possible. A comet can in its ascent and descent approach a planet just as well on one side as on the other if we assume that its orbit is determined by chance. Here the computation of probability [of all satellites orbiting in the same direction] is quickly done. It is as much as when between Cajus and Titius the lot is drawn ten times, which, considered in itself, can favor the one as easily as the other, and when one posits that Cajus should be the lucky one in all ten draws. The improbability in question is 1023 times greater than the probability. Therefore it would be about 1000 times more probable that some of the satellites moved from east to west if their motion was a sheer accident. Another consideration, which makes this improbability even far greater, is given to us by the very small inclination which the orbits of satellites have with respect to those of the planets.[3] Should not one assume here [16] that all comets, which turned into satellites, had but a very small inclination in their own previous orbits and that none of them passed either above or below the planet if this indeed had captured them?

Our telescopes are not yet powerful enough to measure exactly the diameter of the satellites of Jupiter and Saturn,[4] or else I would invite you to consider whether there is not really a certain order among them in the sense that the more distant ones are noticeably larger than the closer ones, in much the same way as is the case with the planets, and the exceptions are unnoticeable.[5] I must leave undecided a similar consideration with respect to their rotation on their axes, which we know only about the moon. But this one example is enough to assume here, too, some purpose and not mere chance. How could it be that of all the comets, which might ever have passed by the earth, only that was caught which turned on its axis in such a way as always to hold its very same side toward us?[6] If this is a chance event then I must say that its probability is smaller than anything one can conceive. The reason for this rotation I cannot find, but infinitely less can it be explained by chance. Where in Whiston's explanation has the moon finally fled during the deluge in order to be safe from a comet whose vapory layer the earth traversed, I cannot [17] explain to you. The moon must have been at least at a considerable distance, or have undergone a great change in its orbit.[7]

You see, Sir, from all this that concerning the many speculations about the comets' effects I estimate very low their possibility, and especially their probability, although I am far from denying all of them; but while I am always willing to grant the minor effects and consider the bigger ones very rare [I think] that one should either consider them as mere exceptions in the doctrine about the perfection of the arrangement of the world-edifice, or [that one should assume that] they in fact serve the purpose of preparing the system of each fixed star to future changes. I hope that you will now find less reason to be angry with the brave Copernicus for having dislocated the earth from its rest, because you see that each comet can do that, and that perhaps several had already produced various changes in that connection. You can also understand from this that we are perhaps not yet Copernican enough, inasmuch as I am not of the opinion that we become such by assuming that the earth can in the long run become the satellite of a comet. I rather believe that comets and planets can, according to the true arrangement of the world-edifice, readily evade one another through entire world-epochs, and that this evasion can always remain possible through such smaller perturbations.

[18] I am also perfectly at one with you in the respect that you, starting from similar grounds, have eliminated from your comet-systems all such features that might bring about disaster as time goes on. I know that your sensitive heart extends its compassion even to whining animals and that you are no friend of upheavals, but wherever you can you build concord, peace and quiet contentment, and thereby make your kindness and friendship so valuable that like-minded souls and friends find in your company the utmost happiness. This you have procured for me to my greatest satisfaction and I shall always feel it in its freshness as I remain with mutual trust, Yours, Sir, etc.

THIRD LETTER

I am beginning to look at the description, which you, Sir, make about the constitution of the world-edifice, with eyes other than [the eyes] I was looking [with] at the description the philosophers had given me. I was not far from looking at astronomers as authorized prophets and, if I am to make matters look even more frightening, from seeing in the invention of the telescope and in the rapid growth of astronomy the herald of an impending disaster. How could, I thought, a genie suggest to Copernicus the structure of the world, to Kepler its laws, and to Newton that terrible attraction and the doctrine about the course and impact of comets, so that everything might be available for the prediction of the calamity and the inhabitants of the earth might see to it that instead of having all come to an end a seed for propagation might remain alive on the changed earth! Could this be the decision of a providence which cares for the preservation of its creatures in all mishaps?

But your letter, Sir, and I thank you for it most obligingly, removes this uneasy prospect as you postpone, though not completely eliminate, such calamities by many centuries. Your picture of the world-edifice indisputably contains something great [20] and something worthy of the Creator's wisdom. You have reflected on the preservation of creatures in a much higher and nobler manner than our philosophers who seem to think only of misfortunes or at least seem to take delight in instilling fear in us.

I have considered with the keenest interest all parts of your letter so as to form myself an idea about the arrangement which you present as the work of the Most Wise. I now strain my imagination and all my thinking ability to follow you along the conclusions which you have presumably carried farther than you wished to tell me, because you say expressly that I have already more than I requested. No, Sir, if I may object, it is because you speak of things other than of derangements and want to introduce among celestial bodies that harmony which is among friends whom an inner drive attracts to one another, even where they see one another [only] from a distance, and where other factors do not permit their coming together. Thus while celestial bodies approach one another, higher commands have it that they continue their journey into the farthest spaces and care is taken for the preservation of their inhabitants.

You present the worlds to yourself in such a way, don't you, Sir? I find in it a true pleasure and I shall no longer look upon the beautiful Jupiter [21] as a rapacious tyrant but as a beloved father, who always keeps his four children around himself, who illumines their night and proceeds with them undisturbed on his peaceful course. Each comet that enters into his domain bows before him and would rather change its orbit than induce a warlike calamity. For such are the thoughts which you, Sir, have offered to me, and I must therefore ask whether all comets, which now orbit from east to west, are not already possessed of such noble obsequiousness toward this prince of planets?

Such a doctrine is at the same time more pleasant and more likely, and I do everything to make the reasons which you have given always clearer to myself. With you I now let each celestial body be what it once was, and I think no longer of the demise of all its inhabitants which would be unavoidable if the displacement in its orbit were excessively great. You are far from giving me enough comfort when you say that I should be prepared for a Siberian and for an even more bitter cold if a comet were to carry away the earth. A winter that would last at least 70 years,—had we been forced to go with the comet of 1759 which still returns most speedily of them all,[1]—such a winter still bears no comparison with the polar regions, and the Dutch sailors, who must spend half a year near the pole, would hardly find anything [22] pleasurable at the return of summer if the earth itself receded from the sun. We are created for the place which the earth really occupies and the earth should remain forever bleak, or a new creation should occur, were it to be carried away.

But I do not now go farther with my fearsome speculations and I shall rather keep to considerations which you, Sir, submitted in connection with the satellites. I have studied them with the utmost zeal. I have looked at a comet from all sides to see whether it could turn into our moon. I have further assumed the most favorable conditions, and since I must give a velocity to the comet which is 27 times slower than that of the earth, I here stipulated the position of full moon to be the aphelion of the comet and I have made the comet move with much the same velocity with which the moon now moves.[2] However, the ellipse, which it previously described, becomes now impossible, for the perihelion point of the ellipse falls now inside the sun. If I move the aphelion farther out, then the comet would pass by the earth either with too great a velocity to be captured by it, or its ellipse becomes even more of an impossibility. I have pursued this consideration in no other than a general fashion. Thus I have also found that either the comet should from the start have found that orbit around the earth [23] which the moon now occupies,—and I cannot see how it could have come into that path because the moon constantly follows it without abandoning it again—or I should have assumed that the whole thing should take place very gradually and that the comet should come only in the course of time into that steady position in which we now see the moon. I cannot, however, reconcile this with the laws of motion, because all depends on whether the comet had at any point of its path a velocity which agrees with the distance of that point from the earth. Once this is so, no change in the orbit is any longer conceivable and the steady position is on hand from the start. In the other case, the comet cannot stay around the earth, and at most its course around the sun is somewhat changed.

In such a way do I present the matter to myself, and I find therefore no possibility of making a satellite from a comet and, as you notice, much less a moon like ours which in so extraordinary fashion and through unfathomable reasons completes in the same time its revolution in its orbit and on its axis. The improbability, which you, Sir, calculated in general for the satellites, makes perfect sense to me, and I am ready to consider it practically a moral certainty, even if you had not afforded other reasons which convince me [24] more than enough. I only wish that instead of frightening us so much, the philosophers had also tried to consider the universe, as well as the smaller creatures, from another aspect and had in a more complete theory shown us the language of heaven, which, I think, would teach us not only the greatness and might of the Creator but also His wisdom and goodness.

How overpowering in that respect is my desire to learn, and how small, on the contrary, the hope to satisfy it in a short time! I have tried everything to make the comparison you have recommended to me and to infer from the countless designs which we find in things on earth those that exist in entire celestial bodies. I repeatedly considered what purpose was served by each part, by each muscle, by each member of our body; why a particular part was in this and not in another place; how perfectly it was adjusted to its end; what is conducive to its preservation; how it was secured from each mishap, or what contributed to its restoration in each injury.[3] I extended this consideration to each animal, to each insect,[4] and I sought how its parts are arranged, so that it may thereby become what it was destined to be in the universe. I followed its fortunes through each season and noticed what changes it undergoes, and how it copes with heat and cold. I even surveyed each variation in the weather and [25] investigated to what end its course must serve in plants and in the animal kingdom. And up to that point Nieuwentüt[5] and Derham[6] were welcome teachers to me. But as soon as I rose above the atmosphere and wanted to find such designs, such arrangement, and such variety on the stage of heaven, I began to be rather astonished and to wonder in quiet reverence, rather than to perceive something more specific. It may be that the order prevailing there is too extensive to be grasped by us, or perhaps it will take centuries before a series of changes unfolds to us, and therefore I readily grant you that here I could not go too far.

Since, however, you have recommended to me the cosmological reasons, I have gathered new strength to apply them regardless of whether I can only hope for general considerations, just as you yourself think. I saw well that I must consider the whole world-edifice as a well-arranged machine. In that case I first had to find general laws, but with each of them also a thousand smaller transformations and ever new applications to other and diverse cases. I took, for example, the light, which all fixed stars have but which in each of them has its special intensity and variation. Then I concluded that the planets and comets illumined by starlight demand the [particular] intensity or its [26] variations. I perused the laws of gravity which are equally general, and hence I found innumerable diversities in the course of planets which are around the fixed stars. The dispute which the philosophers have long ago stirred up about gravity[7] is magnificent and great, and it causes a respectful wonderment if one considers that the whole heavens and all celestial bodies are moved by one and the same law; indeed, if no other law can take its place, this alone must be sufficient to show that through it all the heavens are in a very close tie with one another, that the entire world is an interconnected whole, and that it is not patched together from single, broken pieces. These tenets are undoubtedly great and beautiful, but they still contain so little that is specific.

Since even in such a way I was unable to find much, I turned around and tried to begin with observations, however incomplete they might be. I therefore took Halley's Table[8] of the computed orbits of comets which had been carefully observed up to his time. I compared these with one another in such a way as to see whether they could be brought into a certain order. Altogether I found 24, or 21 to be specific, because one of these comets occurs three times and another twice.[9] Although I could very well see that this number is still very small, and thus in the order which I sought [27] many gaps might still remain, I nevertheless take the view that among these 21 there ought to be more than one kind, because Halley did not choose them but took them as he could have them.

You may ascribe, Sir, this undertaking to an overwhelming curiosity which your letter awakened in me. I will give you now the result and my conclusions which I reached in the matter, and I beg that you further study it and communicate with me if your labors seem to be rewarded. You should always keep in mind that I wished to penetrate the astronomical mysteries, and I would be satisfied if these efforts brought me to their trail on which you must have discovered noteworthy circumstances, as I take it from your letter. I began with the perihelia of these comets and found in all 21 only 2 whose perihelia were farther from the sun than the earth in its mean distance [from the sun], and the difference was insignificant. The remaining 19 had their perihelia between the earth and the sun, in that only 2 passed between the earth and Venus, whereas 11 passed between Venus and Mercury, and 6 between Mercury and the sun.[10] From this I could assume nothing else than that the comets, whose perihelia are farther from the sun and are close to the earth's orbit, are seen more rarely, or that they also return more rarely. [28] For I still should keep in mind that there are comets that never come closer to the sun than [the orbit of] Mars or one of the superior planets.

Then I compared these smallest distances with the angle of inclination, and I found that it was always more than 30 degrees with all the 6 comets that descend below Mercury toward the sun,[11] and over 60 degrees with four of them. As to the remaining that do not descend so deep I found all angles of inclination without exception, but in such a way that the smallest of these angles was above 5 degrees. I know well that one had already concluded long ago from all this that the angles of inclination were greater in the case of comets, so that they could evade the planets.[12] But then why were the comets made out to be so fearsome?

Finally, I found that comets, which occur for the second and third time in the Table, had each time another position in their orbit,[13] [and] although the change is not great, it still shows to me that one cannot yet conclude exactly from the present orbit of a comet its future path.

What do you believe of this, Sir? Could not the comet of 1680, which was depicted to us as [29] the most fearsome, encounter before its return other comets to which it readily adjusted its orbit, so that it no longer passes as close to the earth's orbit as happened in 1680?[14]

You now see my discoveries according to which I would like to arrange my comet-system if you would help me. The closer a comet comes to the sun, the greater I would make its angle of inclination and the smaller angles of inclination should remain excluded. As to the rest which remain more distant I could set the angle of inclination greater or smaller, and I would see to it that a comet should have all the more free space around it, the larger it was. For I see the same also in the case of planets, as Jupiter and Saturn are very far from one another and also from the others.[15] Do you think that I can also posit parabolas and hyperbolas as their paths, or should these only be ellipses? It is demonstrated that all conic sections are equally possible, provided the sun is in their focus. What would you now make of those comets that visit us only once and then take their leave forever? Certainly, I would like to put the comet of 1680 in a hyperbolic orbit if it was supposed to threaten us with a disaster such as one would have us [30] believe. But in your system, Sir, it would be a pity if it would not repeatedly come back, because it is always one of the most beautiful and most worthy to behold.

I expect your reply with the greatest pleasure. How much I wish to be with you once again! I would look neither for ellipses nor for hyperbolas if only I could come to you, and the straight line must always be the road to friendship which leads me directly to you, if this all too slow fate ever grants me that. In this pleasant hope I remain, Sir, etc.

FOURTH LETTER

You come to me, Sir, with your reflections on the arrangement of the world-edifice, and the labor which you intend to spend on seeking support for them from cosmology and experience, and on communicating them to me so graciously, obliges me now to offer my thoughts to see whether I have anything to contribute. Your zeal should serve me as an example, but tell me, is it only a friendly obsequiousness that lets you make no objections to my system, or do you want to be satisfied with the easiest way of banishing fear from the world? You are at least as good a philosopher as those who had conceived those horrors, and the right given us by philosophy to demand for every pretence a proper reason is, I know, familiar to you in all its extent.

I willingly grant you that to me my system is more enlightening only because it seemed to me that it better suited a world which ought to be the most perfect, and on this basis it seemed to me to follow in a most straightforward fashion that evils ought to be the more infrequent the greater they are. I found no reason to assume a new creation, and much less did I want to make the whole world-edifice [32] barren and uninhabited. One entire side of the world would thereby become forever ignored. A planet, which would be derailed from its tracks or forced to follow a comet, appeared to me very close to its destruction and I held all its inhabitants to be without any hope of renewal and propagation, forever lost and wiped out. Transfer the animals that have their habitat around the pole and are equipped accordingly to the sandy deserts of Africa, or animals that live there to the icebergs of the polar regions—such a transfer still says very little. Let the seas dry up, however, and give the fish the air for their habitat, then we come remarkably closer to consequences which the derailment of a planet should entail. But the improbability [of this to happen] will also rise to its true degree.

Undoubtedly, when I posit the course of celestial bodies to be so directed that even the smaller changes which they induce in one another serve the purpose of always evading one another, I assume something which does not lend itself to a rigorous demonstration. I am taking there the wisdom of the Creator in its full extent. I also posit that the preservation of celestial bodies and of their inhabitants is such an aim of the creation which allows no such exception that would include a complete destruction. Once the validity of these propositions is granted, [33] the matter is still more one of wonderment than of complete proof, and to me one can always call into doubt whether such an arrangement, possible in itself, is indeed the case, so that even the preservation of celestial bodies and of their inhabitants can suffer no exception.

You may see, Sir, from all this that I have not made my system free of doubts, although I do not have the intention to let these loom unnecessarily large. Still I must tell you whatever suggests itself in that connection. You shirk, certainly for good reasons, the kind of discourse which includes sheer chance in the processes of the world. Thus, for instance, in connection with the comet of 1680, that passed so close to the orbit of the earth that it was not farther away from the earth than the moon,[1] you are, Sir, already long disposed to avoid the expression: it is sheer luck that the earth was not there at that time. For, indeed, luck and chance are removed here as soon as one considers the matter a consequence of the arrangement of the world, and it will be called a plan of the Creator that care is taken in such a way for the preservation of the earth and of the comets. We should necessarily consider everything that actually happens as means and plans which are in the wisest manner coordinated with one another in the eternal decrees.

[34] Again, should it be possible to prove from the laws of gravity that it is impossible for two celestial bodies to collide, or that should this once happen their orbits would have to be adjusted intentionally to that end because wholly individual circumstances are here required, then in the first case we have no worry about the preservation of entire [celestial] bodies; in the second case, however, the burden of proof would always fall on those who wanted to assert it, because here I simply exclude all chance and can certainly assume that such individual circumstances are substantial exceptions from the general intentions of the Creator. Extensively as we may know the orbit of comets, there are yet seen no such intersections [of orbits] which could ever entail such disruption, and what you, Sir, have remarked about Halley's Table, that comets returning for the second and third time always had a different orbit,[2] can already show that even when such intersections once occurred they did not remain unchanged for long.

The collision of celestial bodies always seems to me to be far removed from the intentions of creation, because each small change in the path of a comet is sufficient to avoid it. And the consideration you offered, namely, that comets cannot become satellites,[3] amply proves that satellites had been satellites from the very start. From this I quickly conclude that each planet, [35] satellite, and comet is in this and not in another place, because it was placed right there and because its inhabitants are fashioned for it. What would we do beyond [the orbit of] Saturn and in a more than seventy-year-long winter?[4] Moreover, such a winter, as you remark, would be still the shortest among all those that comets must endure.

I know, of course, that these considerations which I formulate in great numbers are only fragments of the proof that one can expect for my system. From our brief experiences one cannot yet infer [all] future experiences, and here the cosmological reasons cannot yet be developed to perfection. I wish with you that the philosophers had given some concern to whether the preservation of entire celestial bodies can be reckoned among the Creator's intentions that admit no exception. At least I took pains to make these exceptions so minute that they might be considered infinitely small; and my system about the world-edifice will at most be faulty in that I assume it to be more perfect than it can possibly appear. The extent of the knowledge and wisdom of the Creator is infinite, and I shall guard against setting any limit to it. This would happen if I were to consider a perfection, without the proof of its implying a necessary contradiction, as impossible and too great, or [36] were I to omit in the whole what we admire in the small?

I have perused with great pleasure the considerations which you, Sir, presented in connection with Halley's Table, and I have compared them with the Table. You may proceed with good reasons to fill the space around the sun with ellipses according to the rules which you have set, and I agree in full with the reasons given by you why comets, which do not approach so close to the sun, occur less frequently in the Table. It is truly necessary that comets which descend below Mercury should be seen either in their approach or during their return to the earth, and the only case where this does not very well obtain is the one when the longer part of the ellipse leans downward, for then they are mostly visible only beyond the equator, because they remain for us under the horizon during the best circumstances [of visibility]. The same holds true of those that pass between the sphere of Venus and that of Mercury. They need several months to complete that path around the sun, and have a stronger light from the sun, and for the most part also a conspicuous tail because it elongates itself in approaching the sun as the comet of 1744 has given a clear example of this.[5] The earth may stand at such a time either in opposition to or in conjunction with the comet and the sun, and therefore a few [37] months are enough to see it either in the evening or in the morning. This reason is enough to show why among the 21 comets of Halley's Table there were 17 that had their perihelion within [the orbit of] Venus and 6 within the orbit of] Mercury.

It is quite otherwise, however, with those comets whose perihelion is as far or even farther than the sun from the earth. Their greatest illumination is weaker, their tail is not so big, and the earth must be in a very favorable position if we are to see them, and even [then] their visibility lasts at most a few days or a few weeks. Should they come very close to the earth, their visible course is very fast and each day they recede in the sky by 10, 20 or even more degrees. Therefore in a short time they diminish in size and disappear from sight. And since their smallest distance is still so great, one can easily assume an elongated ellipse and put their return at many centuries away. This again may contribute to making their appearance more infrequent. I do not, however, think that this counts much and [I think that] the former considerations are the more valid reasons. For, since the space around the sun increases with distance, there is place for more ellipses, and more comets can [therefore] reappear [regardless of] whether each of them stays longer away [in faraway spaces].

[38] I reckon in this class almost all comets which one sees only a few nights, though different ones almost every year. The period of their visibility is often too short to determine their orbits from it. Many of them are hidden to us by the clouds and, when one needs telescopes to discover them, it is well to recall that spotting them does not happen too often, but for the most part only occasionally. Moreover, since comets have, in spite of their large atmosphere, a very low brightness, it is not to be expected that we can see them as soon as they descend below Mars, because we soon begin to lose sight of even those visible at that distance from the sun. The comet of 1759 stays almost five years within the circle of Saturn but is visible to us for hardly as many months.

Halley's Table, as you, Sir, remark, gives us only a small start about the plane of cometary orbits; not all kinds of them are in it. Take, for instance, the perihelia. Let us put them yet not farther than the earth [is from the sun], and let us have them remain within this limit, and let us count only once those that occur in that Table twice or thrice. Among all these [perihelia] distances [in that Table] there are 7, and therefore about one third of them all, that fall between the numbers 50,000 and 60,000.[6] I should think on that basis that such is the situation of a comet whose visibility [39] is best assured on our earth. Between 70,000 and 80,000 there is none in the Table[7] in spite of the fact that the space between these two distances is much greater than between the first two. When we also recall that the comet of 1680 comes closest to the sun of all of them, then we can still think of at least five or six others that can come just as close without disturbing one another. So could 12 and even more remain at double that distance, and I have a distinct pleasure in making the number of comets grow as the square of the distance of their perihelia, and extend that proportion rather beyond Saturn. Still I must give you another small example of my model and I will proceed with it with as much caution as is possible for me.

I take, according to your remark on Halley's Table, not more than 6 comets which pass between Mercury and the sun. This is undoubtedly very reasonable. Then, when I am ready to fill the gaps in Halley's Table, a few hundred and even perhaps a thousand [comets] are not enough. But I stay with the 6 which are in the Table. I will not now push the perihelia farther than the orbit of Saturn in spite of the fact that there is plenty of room outside it, since the nearest star is at least 50,000 times farther from our sun than Saturn.[8] You see, accordingly, how gingerly I proceed. When I now redo the calculation [40] I find the square of Saturn's distance to be 600 times greater than that of Mercury. And I do not make that place more full of the perihelia of comets than this. The computation is now quickly done. Thus I will come up with not less than 6 times 600, that is, 3600 comets.[9] Would you think that we had so many neighbors around our sun? Still all these seem to me very few. I then assume that we do not come to see any of those whose perihelia are farther than Mars is from the sun,[10] [noting that] the area of Mars' orbit is 40 times smaller than that of Saturn; and so I find again by calculation that of those 3600 comets only a fortieth part and therefore not more than 90 come into view.

This number is unquestionably small. Just look up a list of comets that have really been seen. Will you not find that more than a few hundred have been seen, not counting aereal phenomena which were lumped with them in former times?[11] You must take at least twice this number because many could not be seen partly because of daylight, many others because of stormy nights, and even more because of the southern latitudes. Thereby you replace more than sufficiently those that have come to us more than once, which does not happen too often because most stay away for many centuries and are not even visible at each return. You have removed even from those 24 which stand [41] in Halley's Table only 3 on this ground, and it is therefore very possible that there move around the sun comets which since the deluge have hardly once been with us. Moreover, not a year would easily go by but that one would not see at least one comet if it were easy to track down those which are less visible and if, in the manner of the Chinese, night watches were to be set up against the sky.[12]

When I take the comet of 1680 as a standard, I could bring all the 3600 comets within the orbit of Mercury alone. For its perihelion was 60 times closer to the sun than the orbit of this planet. Square 60 and you will necessarily get 3600, and according to this measure there would be 600 times 3600, or over two million comets. And since beside the comet of 1680 there could very well be another 2 or 3, I would assume five millions.[13] Do you think, Sir, that this would be too much?

The course of a comet along a parabola as well as along a circle is, considered in itself, just as possible as that along hyperbolas and ellipses, but nature allows no such complete regularity. You know, Sir, how closely these lines border on one another and how little it takes to transform a circle into an ellipse and a parabola into a hyperbola, [42] or even into an ellipse. Assume now that one of the planets would really go around in a circle, then you would easily find that each comet that comes into its vicinity could change its path and make the circle somewhat oblong. The planet would not therefore stay long in that circle. And if Saturn were to move into a circular orbit, its next conjunction with Jupiter would change that orbit into an ellipse. Exactly the same holds true for parabolas. It is physically impossible that they should not turn in a short time into ellipses or hyperbolas.

It is true, that in the case of circles such small changes mean nothing [and] the planet would always remain with the sun even though its elliptical orbit would become now greater now smaller. On the contrary, in the case of parabolas the difference will be more noticeable. Then a parabolic orbit would turn into an ellipse, and the comet stays with our sun and obtains a periodic course. Should, however, a hyperbola arise, then we shall not see it again because it gets necessarily and forever farther away from the sun. I will not assert, though, that there are such orbits; but when there is one, it is necessary that the comet come more and more into the domain of another sun and presumably takes many thousands of years on its way to it. When, however, the comet approaches the fixed star [43] toward which it moves away from our sun, it will obtain around that star—since there, too, the law of gravity holds—an orbit along one of those curved lines named above, and as a result it will stay around that star or set out again toward other fixed stars. The latter eventuality is necessary as long as its path is not turned by some comet or planet into an ellipse, and this change is easier or more difficult in the measure in which the hyperbola, which it began to traverse, is more or less different from the parabola. It is in such a way that one may think of celestial bodies, or comets, which do not stay with any fixed star but are predestined to visit one after another.

I have already often paused to consider such comets and was concerned about their inhabitants. Would you guess, Sir, what kind I picked to be denizens there? As you may very well think, I have simply made astronomers of all of them, created for the purpose of viewing the edifice of the heavens, the position of each sun, the plane and course of their planets, satellites, and comets in their whole interconnectedness. The former would take place during that long period which allows their habitat to go from one sun to another, and the latter would happen when their habitat was on the point of seeking a new path [44] around a sun to observe that sky from a new angle. They would need centuries, just as single hours go past us, and immortality ought to be their heritage because time is allotted to them according to their performance, just as on our earth there are insects whose whole life begins and ends in the course of a few hours because their activity demands no more time.

Don't you believe, Sir, that the world-edifice should also be viewed from this vast side, or should the All-wise, who arranged the worlds, be admired only in smaller things that can be seen by us, and should the arrangement and disposition of all suns and wandering stars remain unexplored? I think the latter [too was to be achieved]. Just as we discover with magnifying glasses in each speck of dust a living world and in each droplet a sea of creatures, these astronomers find the sky full of celestial bodies. And just as a few hours go by in our reflections, thousands of years pass during their considerations of entire solar systems. You know, of course, Sir, that time and space are neither great nor small, but that they should be considered only in their relation to one another, for both become bigger and smaller with one another. A sailor who travels to India has long been accustomed to measure his voyage not in hours [45] but in months, and daily experience teaches us that years can go by like single days and, on the contrary, hours may seem to us longer than days as often as we are forced to wish that they be soon over. So I count with longing each moment, which becomes for me entire years since I have last seen you. The hours shorten during the perusal of your letters, and whenever I think of your company and of the once-so-short hours that passed swiftly during our conversations, I am longing that you come again and that I should not be deprived of the longed-for hours about which I assure you with a thousand joys, Sir, etc.

[46]

FIFTH LETTER

Your world-edifice, Sir, is too pleasing to me to let myself make objections to it too lightly, and when some come to my mind I would much rather submit them to you as questions which might serve you for its further explanation. But I am not concerned about this. You certainly must have further developed your thoughts, as you noted when you gave me to read your writings which came forth on the subject. And finally, to consider the matter properly, what objections should I make to you? Against your main purpose? This would not occur to me even in my wildest dreams. It consists in your resolve to find in the whole world-edifice the kind of order, harmony, variety, transformation, interconnection, perfection, beauty, means, and purpose which we admire on the earth even in the smallest things. You seek to make the exceptions infinitely small and even to call into doubt whether exceptions occur, or whether they should not rather be considered as a means through which transformations become more variegated and the order in the course of celestial bodies becomes more perfect and enduring. You ban everything that comes close to blind chance, and what we often, without reflection, call luck, you turn into a result of the arrangement of the world and [47] into a proof of the wisdom and goodness of the Creator who chose and created the most perfect of all worlds.[1]

You always maintain that it takes world-ages until the consequences of this arrangement become clear and give as evidence that the world was not created for a few moments, as are worms that require only a few hours to bring about on the surface of our earth the brief changes to which they are destined, and for which all proportions of their bodily parts are arranged. What you, Sir, are saying of entire world-systems is suited to their greatness, their splendor, their duration, and their destination, and the doubt which you yourself raise about the possibility of such an arrangement vanishes when one reflects on the infinite in relation to the omniscience, power, wisdom, and goodness of the Creator.

Why do you want to continue with doubts? Is it not that the principal question hinges on whether comets and planets might forever evade one another? Do you call for my objections? I from the very start frowned on the philosophers because they supported themselves with the flimsiest reasons, if this was possible at all, and they would have had us believe that the world is so arranged that before long we should [collide] with a comet. But you, Sir, take the most likely of all possibilities, because they are of them all the wisest and [48] the best that were [available] in the treasures of God's omniscience. Do you hold fast to the great tenet? The whole world is a continued effect of all divine perfections taken together, [and] what else would you derive from them than love, goodness, infinite power, wisdom, care and preservation of the whole, order, duration, and perfection? Would you not necessarily conclude that the duration of the whole should last forever, and the preservation of each part should come the closer to eternity, the greater they are and the closer they come to the whole? Is it not so that whatever is mortal propagates its species, and what is subject to change renews itself? Where do you find exceptions to these eternal laws, and if you find none, where would you look for apparent grounds for the collapse of world-systems? You use in your world-edifice the law of gravity which some genie revealed to Newton in a moment of rapture, and the paths of celestial bodies are thereby demonstrated not arbitrarily but in a straightforward and irrefutable manner. Should I dream about more improbable circumstances to investigate whether a comet could become a satellite, and to make ten comets move exactly so that finally they enter around Saturn, Jupiter, and our earth into such orbits which differ as little from circles as do the orbits of the planets themselves? No, Sir, think not that I would go out of my way to that extent [49] in seeking objections and contest the most probable with the most improbable. I simply keep maintaining that satellites have always been satellites, and the similarity between their course and the course of the planets in respect to direction, as well as to inclination and to the roundness of their orbits, shows me always more of the planning which is the intent of the great Creator and not the fruit of chance and confusion.

The more your world-edifice unfolds to my understanding, the smaller becomes the difference between my cosmology and yours, and I now foresee that I soon will cease complaining to myself about the generality of the grounds of that science. I only wish to have still more Tables similar to that of Halley. How willingly would I send them to you, and what important remarks and comparisons would I expect about them from you! But tell me, Sir, do you seriously hope to produce still more than 5 million comets? For you gave me this calculation only as a small specimen of your project. I confess to you that I was rather taken aback at this point in your letter. Are you here not too generous, and what do you want to make of such an enormous number of comets? You know that many of our philosophers held them to be immature planets, and doubted whether as comets they could be inhabited.[2] Take [50] the comet of 1680, for example. On December 8 of that same year it was 160 times closer to the sun than the earth. This increases the heat coming from the sun 25,600 times, since it increases inversely as the square of distance. Assume now that the best burning glass can amplify the rays of the sun 2000 times, which may be just too much or too little, then it can readily be calculated that the heat focused at a point by 12 of the best burning glasses is hardly as great as the one to which that comet was exposed. I prefer to make the calculation in such a way because it is more correct than when I compare the overheated comet with a glowing iron, or state that it would not cool off during the next 50,000 years.[3] I will not even [attempt to] determine the heat which the comet would really have, but I know that I would rather endure with the comet of 1759 a seventy-year-long winter than such a heat with the comet of 1680. The philosophers wanted to derail such an extinct comet from its orbit in the course of time and to make it a planet, and in exactly that way does Whiston present us the earth when it was still a chaos.[4]

For what purpose would you have such a rich store of immature planets which are yet to cool off? But I will be thoughtful. In your world-edifice you let each body be what it has originally been, and you grant that comets can turn into satellites [51] just as little as you would want to make planets out of them; you avoid thinking about new creations and you do not even want to leave the comets uninhabited. I must ask you again whether you do not exaggerate the matter and thereby make your system improbable? You turn the whole world around and as you want to diminish hostility in the firmament, you get involved in a clash on earth with the philosophers. Certainly, Ovid has not come up with a more extraordinary metamorphosis[5] than the one to which the comets are subjected in the hands of the philosophers. Aristotle took them for meteors.[6] Kepler, Hevelius, and others made heavenly clouds out of them.[7] Until then everything was transitory with comets. Soon, however, the comets began to vie with the planets in dignity and longevity of origin, and indeed rose to the dignity of celestial bodies and asserted this right of theirs with impressive reasons. But can I give you, Sir, unconditional support when you make the comets the chief part of the whole solar system? How far do the planets fall from their previous rank, and how little can they, whose number when counting also the satellites is only 16,[8] be set up against legions of comets? Where does our earth remain which formerly as a queen high on her throne viewed the sun and the moon as her two lamps,[9] let the planets and fixed stars wander around her as satellites, and granted the comets but a brief sojourn in her atmosphere?

[52] Now, Sir, you will no longer make the objection that I take no exception whatever to your world-edifice. Still perhaps I was somewhat too hasty. I know all too well that you do not assert paradoxical propositions without good reasons. However, I cannot here discuss all my questions, and I will ask you for their solutions. How pleasant it is for me that here, too, I can pose you a problem so that your reasons may become clear to me! Look now, where I find the problem. Do you believe that comets and planets are bodies of different kind, or that regardless of the difference in their appearance and orbit they belong in one class, not only inasmuch as they are neighbors and move around one sun but also in other specific regards? If you, Sir, assume the first, then I must ask again, why do you believe that there are so few planets and, on the contrary, so many comets? This question seems to me the most difficult.

I will now allow a few hundred, or if you wish, a thousand comets, and should the matter depend on my pleasure I would make room for your sake for 5 million without further ado, and I am not going to be taken aback by the result that thereby the whole space around the sun would be filled not only with the bodies of comets but especially with their atmospheres and tails. For I grant you that both these become [53] greater only when the comets come close to the sun, as this was plainly observed with the comets of 1680 and 1744, and with others.

But to show you that I have investigated also the reasons supporting this choice, I have done separately four calculations in an orderly manner. I first wondered why you did not let the number of comets increase with the cube of distance of their perihelia from the sun although you had their increase so much on your mind. You know well that perihelia can have all possible positions in each spherical space that can be imagined around the sun. Assume now these positions to be distributed in uniform manner, and their number unquestionably increases as the cube of distance. But I soon noticed that you have also been intent on avoiding the intersection of orbits, so that you could give each comet a peaceful orbit. My idea would be well founded if all comets were at the same time at their perihelia. Since, however, they orbit around the sun, many perihelia, and with these as many orbits, are discounted. You do not see each orbit here as a geometrical line, but you take only the strongest part of the sphere of influence of the comet into which no other comet must penetrate. In that case you will correctly fall back on the square of distances.[10] Here I picture for myself spokes which are directed as radii toward one another. There are only 12 which come right [54] to the center.[11] Since, however, all diverge from it, there remain spaces for 12 others to be placed in between. These again leave new spaces for others to follow, which, however, stay farther and farther away from the center. Their number will increase only as the square of distance from the center and not as the cube.

So far I found no fault with your reasons but, if you, Sir, allow me to say so, you prove thereby only the possibility that there might be so many comets. Are there, however, in fact so many? You can derive this from no other than cosmological reasons, and when you wish to assert it seriously, then I will take your opinion rather lightly. Obviously, you have to push the horror which nature has for vacuum so far that you put in the same class empty and uninhabited space, and you simply ban both. You have already asserted to me that even with the best magnifying glasses we see in a waterdrop only the whales and on a speck of dust only the elephants. In such a way I grasp very well that you would possibly leave in the whole firmament no orbit into which you would not put a celestial body, which [in turn] would have on each speck of dust a whole world and in each droplet a sea of creatures. You make the great and bold conclusion: either the earth alone is inhabited, or in each point of the whole [55] world-edifice there are inhabitants and creatures.[12] Since I grant you this conclusion, you may see that it is not up to my mere preference [any more] to accept [or not] your world-edifice. But before going farther in the matter, I beg you once more for the solution of my question on the comparison between comets and planets.

Concerning the remarks which you, Sir, have made on the hyperbolic orbits of comets, I find not the slightest difficulty. I now perfectly understand that your intention is to determine in your system all that our too brief experience leaves undecided by considering the purposes of creation. Here again you use a proposition which deserves to be clarified extensively. You request that the world should be considered in the full extent and connection of [its] creatures in much the same way that we consider on a small scale what is on the earth. Our eyes are directed to rather small objects, and we have enough light from the sun to see them clearly. We have hardly begun to use magnifying glasses and telescopes, and what was disclosed to our eyes is only for those few who have all the necessary inspiration and whom an inner drive leads to explore the smallest and largest worlds, the elements, and the whole. The whole world-edifice and its arrangement practically remain hidden to us. Should it, however, therefore remain a mystery to all creatures, [56] or are its order and perfection not so beautiful, not so worthy of admiration and worship as in the infinitely smaller objects of our senses? Does the earth exhaust the richness of the omniscience, omnipotence, wisdom, and goodness of the Creator of all things, who calls the earth His footstool, but calls the skies His throne?[13] The Highest shows us that He is infinite and great also in the small, and exactly this will be found also by the inhabitants of other planets in an ever variable manner. But one comes no less closer to His infinity when one learns of Him from the basic plan of the whole world-edifice. I at least readily grant that I have not known what space, size, distance, and circumference mean in the work of the Almighty, before I learned to take the path of light for a yardstick and ban the misconceptions of childhood, which raise the stars hardly above the clouds and make the sky stretch above the earth to the extent of a few miles. What an effort, how much resolve does it require to force our way from the narrow confines into which our senses enclose us to the limits of the solar system, and how much do we fall behind even there!

I rank highest the astronomers which you, Sir, place on such celestial bodies. Their route proceeds from suns to suns as we go from city to city on earth, and when in our case [57] a few days go by they count in myriads of our years. They are destined to admire the ground plan of the world-edifice, and understand in its foundation and order the series of divine counsels about its structure. Our greatest measures are their differentials [infinitesimals], and our millions can hardly suffice as their table of multiplication. They know the warmth and brightness of each sun, and with a single conclusion they determine the general characteristics of the inhabitants of each planet which orbits around it at a given distance. Their year is the time [which it takes to go] from one sun to another. Their winter falls in the middle of the intervening space, or of the journey which they make to another sun, and they celebrate the moment when their former course turns into a new one. The perihelion of each course is their summer. Their habitat is suited for each distance from the suns, and each degree of heat conditions their habitat for the growth of such plants that serve for them as specimens of those which occur on planets and comets at a similar distance from the suns. Each entry into a new solar system is their spring, and they celebrate the fall when they leave it again.[14]

Still I must break off because I can, as well as you, Sir, stay too long in contemplation of these celestial bodies. It gives me no small pleasure that there are such creatures who take in at a glance the world-edifice in its entirety, and I wished in due course to journey with these astronomers around millions [58] of suns. How rich would I thereby become in the knowledge of the world-edifice! But this wish remains unfulfilled and, satisfied with my present condition, I will untiringly contemplate it from the distance. The series of conclusions which you, Sir, have made has already brought close those great bodies, [and] you know that I follow whither you advance. Show me the new paths which you have opened up in the regions of the heavens. I will wander with you and say to you at each step that you oblige me in the utmost. Sir, etc.

[59]

SIXTH LETTER

The cosmological reasons which you, Sir, have given me in support of my system are all the more pleasing to me, because I was just about to explore it from this angle in a more accurate manner. I only regret that you have stopped short so soon. Do you rely perhaps so much on the similarity of our ways of thinking that you believe that I must necessarily have in individual details the same conceptions? While I did my utmost each time to approach this perfect accord, I find all the time that other subjects make other and still not foreseen impressions, and in connection with this diversity there arises the satisfaction that it is enough when friends are at one in basic principles, because they then communicate to one another ever new consequences from those principles and they will still be found in consonance. I find, therefore, very agreeable the remarks in your letter, and I hope that you find in the same way my answers to the questions which you have submitted to me. We strive forever to become darlings of the noble truth which is suggested to us by its principles, and differently as we were thinking at the start we shall find ourselves closer together following the investigation of the differences, and each new discovery will serve as a firm ground for new harmonious [60] thoughts. How pleasing is to me the prospect that nothing can separate our thoughts! I can ascribe to this touching harmony that you, Sir, have settled aforehand on grounds which I use in my system. You set them forth to me as elaborate truths which extend across the whole world-edifice and you give me the prompting to explore in all rigor how far they can be tied to one another, so that I may build on them and advance further. I must, however, tell you how I wish to sift them.

I viewed the world-edifice as fashioned after innumerable general and special laws and intentions, and this seemed to me to be the required concept about the highest perfection which the world should have. The

80

most general of these laws I saw as the principal intentions which admitted no exception, and the more special ones had to be enclosed by these. As I endeavored to consider the world-edifice as a whole, laws which admitted exceptions vanished from my considerations, and I saw well that the more special they were, the less they were to be used. I saw just as well that all those that admitted no exception were indispensable to me. This was my project, and you will easily notice that the first question in this connection was how one can recognize such exception-free laws and where one should find them.

[61] For me this difficult question gradually let itself resolve into various others. In the world-edifice there had to be only similarities, but with each similarity also uncounted differences and varieties. These differences I had to put aside and make the similarities so much more universal until I could apply them to each celestial body. What else could there remain for me but the most abstract notions, as I had to leave out all that was individual. Nevertheless, those abstract notions had to be of a different kind, so that in spite of this difference they might be connected with one another, and to that extent I thought to assume for each a general law that had no exceptions.

Thus I considered, for instance, motion and extended the law of gravity to all celestial bodies without exception, because this, as you, Sir, have remarked in your previous letter, makes the world-edifice into an interconnected whole. It is not yet proven that this law is necessary [to assume] as soon as one has bodies in mind, and we do not know yet well enough the nature of the material substance, of which the real world consists, to decide if no other law could have been given to that substance. It should suffice that I take the world as it really is, and I conclude with you that this [world] would be a patchwork if no general law connected its parts together. Newton, who taught us this connection, took pains to investigate still other [62] [possible] laws of gravity and to show their dissonances.[1] Spiral lines, along which the planets would always fly away from their suns or ultimately would fall into them, are consequences [examples] of this. The choice of the All-wise fell on the most simple [law] which had both eternal harmony and order, and through which each celestial body could remain that to which all its inhabitants were destined.

In a similar fashion I have scrutinized the habitability of the world, and I certainly made the bold decision that I wanted to leave no space of it either empty or uninhabited. The world had to have an imprint, or as you, Sir, call it, an enduring effect of all divine perfections taken together. Could I then leave unutilized in this connection a viewpoint from which one can contemplate all these perfections, or could the world be the effect of an infinitely efficient Creator unless life and activity, thought and desire were present in creatures at each point of it? Had I made perfection consist in a permanent and inexhaustible alternation of similarities, but had I left empty places where nothing of the kind went on, where no parts of a whole were present, would the infinite then have been complete? I could not admit such gaps and I did not mind filling each solar system so much with habitable celestial bodies as the splendid order, which is introduced in their orbits, [63] would ever permit. On our earth, which we can view since the invention of magnifying glasses even in the smallest details, we find everything so full of inhabitants that we cannot hesitate any longer to consider filling each part of the world with people and life as an intention of the Creator which allows no exception. This is now taught us on a small scale by the evidence of the eyes, and the degrees by which we advance in improving magnifying glasses permit us the secure conclusion that we have not yet by far discovered the smallest creatures. Should we then set narrow limits to that law when we want to extend it to the number of celestial bodies?

After motion and habitability I compared time and space, and the whole world served me as an example that both increase together and that the small changes replace through their sum only that which in single bigger changes is merely simple. A small change can be quickly brought about, but the more often it recurs further changed, the sum of its varieties becomes considerable. Thus the dust turns over under our steps, and the small worlds of which it is made are transformed into new ones. Should, however, cities be built and forests be planted, then years are needed, and the time during which celestial bodies move around the sun increases with their space to many centuries. It will still become a thousand times greater [64] when a [celestial] body journeys from one sun to another, and no longer allows itself to be expressed conveniently in our years when we want to determine the changes of entire solar systems.

About the condition of the inhabitants of each celestial body I was not yet concerned because I could assume in general that each of them is adjusted to the place where he finds himself. What we find on the earth conforms without exception to this law. We find animals some of which are made for the polar regions, others for the torrid zones, for the highest Alps, and still others for the depths of the earth. Each finds in its habitat its proper measure of warmth, air, and nourishment, and each is organized accordingly as to its bodily parts. This consideration becomes more obvious when we compare earth, water, and air with one another.[2] Who would think of the habitability of water if fish and other aquatic animals had not been known to us since childhood? We would hold it impossible on grounds similar to those for which we deny habitability to fire, and the difference would arise only between being burnt and drowned. It is true that fire is not as widespread on the surface of our earth as water [and] therefore it should part with its right to permanent residence. If, however, fire has its steady place below the surface of the earth, as asserted by many naturalists,[3] this objection would be removed. Perhaps the inhabitants of fire [65] are invisible to our eyes, and the efficiency of fire in the dissolution of materials known to us demands either bodies made of asbestos, or such bodies that do not let themselves be split any further. But I won't tarry on this point. The structure of such creatures may remain forever hidden to us, but their possibility should not be rejected lightly.

The calculation which you, Sir, made about the comet of 1680 is just as clear as the one which is generally given about it, and in respect to correctness it goes incomparably farther. It often made me wonder that the common[ly accepted calculation] which one finds in so many writings is copied so readily and without further inquiry, and in all likelihood this would happen less frequently if the authority of Newton, who gave us only a casual computation,[4] would not serve as a substitute for reason. About the [degree of] heat, which the comet had really reached, nothing whatsoever can be decided, and all that one can do in that connection is to determine the density of the sun's rays which penetrate the atmosphere of the comet, and there your comparison with burning glasses is justified and understandable. Were our earth to take that route and from now on were it to move into such proximity to the sun, it is indisputable that the effect would not be much different from that which would be produced by 12 of the best burning mirrors taken together. The whole sea would [66] turn into a high and thick atmosphere of vapors, and although it would considerably hold off and weaken the rays of the sun, I still would not guarantee that this cover would be enough to shield the earth's surface. So much is certain that the light of the sun, when it is on the horizon, becomes well over 2000 times weaker because of our present atmosphere. The atmosphere of the comet of 1744 was about 8000 miles high. It can therefore intercept and reflect still many 1000 times more light. The heat of a glowing iron cannot be determined so easily, although it cannot be much more than 4 times greater than the summer heat of the earth. In that computation Newton had not yet put deep enough the degree of absolute cold.[5] A body that needs 50,000 years[6] to cool off might have needed just as many years to reach that heat. Finally, each body is capable only of a given degree of heat. The comet of 1680 had therefore to possess quite a special nature, different from all bodies of our earth, when it had to be 2000 times hotter than glowing iron, because this, too, cannot become more than glowing and turns into glass and ashes in case of greater heat. Either its atmosphere had shielded that comet from the sun's heat,[7] or its structure was proportioned to that heat as it passed by the sun, and I do not doubt that its inhabitants were to remain unharmed. Perhaps they are [67] of such constitution that freezing and heat make no impression on them. Only the inhabitants of planets, and therefore we ourselves, too, are so delicately adjusted that we need a moderate warmth. We did not become fit to make the journey of this and other comets. Should I, however, make a comparison, then the inhabitants of planets are compared with those of comets much as plants under the equator are compared with those of the northern zones. These endure all changes of the weather, whereas those must be cared for in warm greenhouses if they are to survive in our lands.

The question, which you, Sir, posed to me on the difference of planets and comets, I cannot as yet completely resolve, though I have worked out various considerations that can help toward the solution. Once I have posited that both should be inhabited and remain what they were from the start, this chief difference simply vanishes in my system. I permit here nothing immature and much less would I grant a storehouse of underdeveloped planets. The external appearance, which in our eyes seems to make a considerable difference, consists in their greater vapory layers and in their tails which we see differently in the planets and in their satellites. But you ought not to be taken aback by this, because vapory layers and tails become greater only [68] when they come closer to the sun. I do not think that this increase is only an incidental effect of the sun's heat. The vapory layer may have an essential purpose and conceivably serves as a shield against an all-too-great heat. Our earth may need the clouds in a similar respect, although these are by far not so necessary to lessen the heat. In a more special manner the earth's surface seems to need a shield against the cold of winter, and the snow serves for it as a cover. Since the vapory layer of comets is larger only for a few months, I will not at all consider the condition of their inhabitants as one enveloped in steady mist and vapor, which would prevent them from clearly contemplating the world-edifice. Our atmosphere never becomes so big as that of the comets can become; it rather remains steady and for half of our lifetime the sky is covered for us with clouds. In a similar manner does the snow cover for us the earth's surface in winter time. But do thereby the heaven and earth remain unknown to us if we are willing to use the time when both uncover themselves again?

This swelling up of the comet's atmosphere is therefore to be looked upon as a consequence of its elliptical orbit and may be necessary with regard to its shielding. But because of this planets and comets are not yet different otherwise than with respect to their orbit. With the former it is almost circular, [69] with the latter very oblong. The question also bears on the possibility whether these circles and long ellipses do not approach one another through innumerable degrees. This question cannot be easily determined through experience [observations]. We are so close to the sun that we are given to see only the interior part of its whole system. Beyond

Saturn we spot no further planet, and such a planet should necessarily remain invisible if it were considerably smaller than Saturn.[8] Of the comets we also see only those which descend below Mars, and the most visible ones must come half as much closer to the sun than the earth, as I have already noted in my last letter. It is therefore conceivable that of all members of our solar system we see only the two extremes, namely, those whose orbit is either the most round or the most oblong.[9] Inside [the orbit of] Mercury no comet comes whose orbit would not be very oblong, although the opposite, considered in itself, is not impossible. The reason for this is that comets, in general, stay away [from the vicinity of the sun] for a very long time.[10] The one from 1759 still seems of them all to come back most rapidly and yet it takes 75 years.[11] When you square this number and from the square extract the cube root, then you find the mean distance of a comet from the sun that returns in 75 years. The computation shows that this mean distance is $17\frac{3}{4}$ times farther from the sun than the earth. The double of that number, or $35\frac{1}{2}$, gives the major [70] axis of its ellipse. How narrow that ellipse must then be when the perihelion is closer [to the sun] than Mercury, and how much narrower when the comet stays away even longer! Even when a comet, whose perihelion is as far from the sun as the earth is, comes back in 75 years, its orbit still would be three times longer than wide. But it is possible that such comets need centuries for their return.

You see, Sir, that this demonstration is derived from experience and from the law of gravity, and that therefore I can necessarily consider the orbits of comets visible to us as very oblong. Of all orbits they are therefore also most different from circles, and I can assume that only the two extremes come into our view. Remove, however, the perihelia of comets farther from the sun, then the ellipses can become somewhat rounder, and I would not set the orbit of the outermost comet very different from a circle. Great as the space around the sun and its sphere of influence may be, the latter has its limits with respect to comets which must orbit around the sun. I cannot remove any of them so far that it may come close to the sphere of influence of another sun. Its gravitation toward our sun must always remain strong enough, so that the gravitation which it has toward each fixed star might be considered negligible. I will not even determine whether it should [71] remain 100 or 1000 times closer to our sun than the [nearest] fixed stars. The nearest of these stars may be 500,000 times farther than the earth is from the sun.[12] Give to the outermost comet only a 1000th part of this distance, then it would still be 500 times farther from the sun than the earth. Should it orbit in a circle, then this distance would be its half axis, and the time of its orbiting would stretch to 11,180 years and would hardly be $\frac{2}{3}$ times shorter when such a comet would descend to within the orbit of Mercury.

From these reasons it may be assumed that beyond Saturn there are still orbits which do not differ much from a circular orbit.[13] But according to my system there should be very few of these, and the oblong ellipses have with me an advantage which must necessarily set the number of planets very low. This is the question which you have submitted to me, and this should also be the last answer that I can find for it. Judge it for yourself, Sir, to what extent it brings you satisfaction. I derive it from the two propositions of my system which you have granted me. The number of celestial bodies that orbit around our sun must be as big as possible, and in their courses they must always be able to evade one another. You know, Sir, that I consider this latter law absolutely necessary, and the number and constitution of comets and planets [72] must determine itself accordingly. You yourself have remarked to me that on this basis I let their number grow not as the cube but as the square of the perihelion distance. Now I will prove why the circular orbits are the most inconvenient.

The law of gravity implies it as a necessary consequence that when comets or planets must orbit around the sun in a circle, the sun must necessarily be in the center of the circle. If therefore you assume that all comets must orbit in circles around the sun, then these circles must be concentric. Since comets must also evade one another,—you may remind yourself from your letter that these circles should not be regarded as geometrical lines,—one must take in the case of each the greater part of the sphere of influence of the comet. Transform, accordingly, your spokes into circles of such size that one should fit into the other and that all remain concentric. Your system will look approximately as an armillary sphere, with the difference that you cannot accommodate more equally great circles. How does now the number of circles increase? Would it not increase simply as their distance from the common center, as you place around each circle the next bigger one? Here it is much the same whether you place the circles in one plane, or whether you give them different angles of inclination with respect to one another. Since they must remain concentric, [73] the neighboring ones will touch one another at two points, and the space which remains empty is here completely superfluous. If you count now only 6 comets within Mercury, then you will come up with hardly 150 up to Saturn.[14] However, I came up with 3600. The difference is certainly very considerable.

Here the chief obstacle lies in that all circles must be concentric, and this obstacle is all the more absent with ellipses, the more oblong they are. The sun is in their focus, and no sooner does a comet move away from its perihelion than its distance from the sun also increases and leaves enough room for a new perihelion. The picture which you have drawn with straight spokes comes much closer to such ellipses, and when you transform the spokes into such ellipses, you will lay around incomparably many more of them than when you make concentric circles of them.

From this, Sir, you see how your system comes so close to nature. I have previously proved that the orbit of comets that approach the sun inside the sphere of Mercury is very oblong, and it becomes clear from these considerations that it was the intention of the Creator to make the solar system as perfect as it ever [74] was possible. Therefore we have so few planets and, on the contrary, legions of comets, and I draw from this the conclusion that both belong in one class and that the comets constitute, though not the most prominent, but still the most numerous and considerable part of the solar system which is visible to us. I would much rather ask why there are still some planets around the sun. If this were not merely so because there remains so little room between the ellipses which the planets describe, then I would simply look for the reason in that around the sun there ought to be also such denizens, who needed an always similarly apportioned warmth. With us delicate beings it had to happen that the earth should move in an almost circular orbit. I think, however, that both reasons concur here because diversity and habitability must go together in the world-edifice. And one of these aims can serve to determine the other. Perhaps herein also lies the reason why the orbits of planets all lie in almost the same plane. For this would leave for the comets free space above and below that plane. The intersections of comets and of planetary orbits become thereby simpler, and from this the reason becomes clear why the planets are so distant from one another, [namely], because between their orbits there had to remain space for all the intersections of cometary orbits. And since these intersections are farther from the sun in the case of very oblong and steeply inclined ellipses, [75] it then follows from this why Saturn, Jupiter, and Mars are the farthest removed from one another. Were still some planets beyond Saturn, their [mutual] distance would be in all likelihood still much greater because there the number of intersections of cometary orbits would be even higher.[15]

These are the reflections which I made à propos the solution of your question. Tell me, Sir, what do you think about it and to what extent are they satisfactory to you. It would be for me very pleasant if you gave me a new opportunity to find more coherence in my system, because what I have already found in that respect I owe it to your remarks which in every case have been useful and obliging. I know that you are inexhaustible when you think of means of obliging your friends, and how many proofs of this you have given especially to me! I beg you to consider my proofs, insofar as they depend on my powers, as effects and signs of the indestructible friendship which unites our hearts even from a distance and with which I remain, Sir, etc.

SEVENTH LETTER

It is obvious, Sir, that your world-edifice will be found verified in a long series of future observations, because you take together all sources out of which there can but freely flow the reasons for building it up and supporting it further. You search according to the rules of perfection which permit no exception, for you necessarily and correctly assume that the world in every case is the most perfect because it is the work of the All-wise. You have determined the properties and distinctive marks of these rules and laws in order to spot them in the real world. General means you make into general purposes, and you give to each circumstance all possible varieties. You extend the power, order, life, efficiency, wisdom, and goodness of the great Creator through the whole world-edifice, because you can see it in the smallest details as the effect of all those perfections taken together. How complete, how harmonious will everything become through considerations that until now one labored to assert only timidly and partially, and how richly do you fill the gaps of proofs that hardly cover single parts! Should it then be called boldness when you seek to set forth such broken pieces, such hardly ventured proofs in their true completeness, or to make truly universal what should necessarily be universal?

[77] It is true, the consequences of your reasons seem to turn the whole solar system around, but this change takes place only in our mind. In the same way Copernicus turned it around previously, or rather, he made the beginning to that end, but you, Sir, seem to complete it, or at least you seem to build the road, so that it may be accomplished through future observations. For it is on these that the determination of the path of each comet will depend; you, Sir, are satisfied with coming forth in these latest times with an art of prophesying and with showing us the ground plan and the general [structure] in the whole system in order to predict what one shall unfold in the future. The meager supply of our observations gives you but very small and fragmentary pieces; you compare these, you discover missing parts and make the conclusion as to how the whole ought to look. The harmony which rules in your thoughts, [and] which is my satisfaction and the tie of our friendship, leads you to the ordering of the world-edifice and expands there through all parts.

What could I deny in the bold and far-sighted conclusions which you make? The sun should give light and heat to the greatest possible number of celestial bodies and these should always evade one another, and therefore there ought to be innumerably more comets than planets. Should I doubt the premises, or state that the sun [78] is permitted to waste its light and heat in a useless manner? It is enough for me that at least as much light comes forth as is necessary, so that the inhabitants of other solar systems may also see that our sun, and with it our whole system, too, is in the world. The view of the starry sky and especially of Sirius is to me too dear that I should deprive its inhabitants of the beautiful brightness of our sun, which to them shines as another Sirius through their nights. Take away from me, Sir, anything but the starry sky and I will give you as many celestial bodies around our sun as you wish. I find the earth much too inhabited on the small scale so as not to be willing to assume as many large dwelling places. Just as a swarm of busy bees swings around their queen, in the same way, and a million times more, should celestial bodies go, as far as I am concerned, around the sun. They go peacefully and united, and none of them should disturb in a hostile way the course of others, for they are to be preserved together.

I have gone through the proof of this proposition in great detail, because it was to be the solution of a question which I held to be the most difficult in my last letter. I took my spokes and bent them into a circle. But I could not make these immediately concentric and of one size, without making incisions on them and then joining them, as the equator and the colures.[1] [79] are joined to one another on the armillary sphere. This meant nothing less than to assume comets whose orbits cut through one another. It is true, the periods of such comets would be equally great, but they would not remain so too long because there always occur small displacements in the orbits, and moreover they would become too uniform. In the world all possible varieties which are permitted by general laws ought to be realized, because perfection becomes thereby greater.

However, I bent my spokes into ellipses and made some of them equally big. Their focus I made into a common center. Some of them I let go downwards, some upwards, and still some others diagonally, and they were intersecting with one another in an orderly fashion. Each went in a different direction and they [all] spread out divergently from the center. There always remained greater space for subsequent ellipses which could intersect in an orderly way with the first ones.

I could still have found between these ellipses plenty of room for a few circles or round ovals, but it always appeared to me as if a few longer ellipses had to be left out for the benefit of the former. But I have not pursued the investigation that far, because I could think just as well that my intersecting ellipses still could be infinitely removed from being a sample [80] of the universe.

I, however, do not believe that you, Sir, leave a place in the world for the planets only on the basis of toleration. I would not in the least sacrifice for our earth's sake some of the five million comets. The loss would [in a sense] be so unnoticeable. Still in all likelihood you have so decreased the privilege, which the planets had in former times, only that you may show that one should look upon the comets from a higher and more important viewpoint and not as immature, or quite antiquated and worn-out planets. For, according to your system, they are useful for the habitability of the universe and for its being filled with people in a more excellent manner than are the planets.

The conclusion of your letter shows me that you could have still drawn a good amount of inferences from your proposition about the number and comparison of planets and comets. But you have satisfied yourself with piling a few on one another and with presenting them briefly. You may have thought in that connection of the *sapienti pauca* [a few words are enough for the wise], but I must confess to you that I would have preferred that you had stayed a little longer with that topic. Perhaps you have set it aside for a more convenient time, or are you perhaps confident that I have made as marked a progress in that science as I hoped to do? In the first case, I shall eagerly wait for the elaboration, and what I [81] will now do in the second case should only serve to show you that I consider your conclusions worthy of all attention and that I try, as is the case with all your thoughts, to adopt them. Would you see, Sir, how far I have succeeded?

You conclude, for example, that all orbits of the planets must lie in about the same plane, because it is clear in itself that they will never perfectly coincide [in one plane] because of the small perturbations.[2] You derive the reason for this situation unquestionably from the fact that you can thereby bring a much greater number of comets into the solar system. This, for you, must be as habitable as possible. I started, however, with assuming the opposite and gave each planetary orbit a steep plane. I left the earth in the ecliptic [and I let] the orbits of Mercury and Venus intersect one another at a right angle at the pole of the zodiac. The remaining orbits I made lean against these under different angles. Then I considered how the ellipses, which the comets describe, could be drawn between the more round ellipses of planets. For the six planets I had already six planes into which I could not place any comet that could come closer to the sun than the planet which entered in that plane. Saturn and Jupiter deprived me of even more comets, because I wanted to place more than one elongated ellipse in the same plane, just as the planets are now together in the [82] plane of the zodiac even though they are somewhat inclined with respect to one another. I therefore soon began to bring again the circles of planets closer to the ecliptic, because in this way all the six no longer impeded me as each of them did before, and I found that even the planes, which you have made for me useless for comets, now could be occupied with the ellipses of comets. Of these ellipses, which I put in one plane, the inner ones had to be elongated, so that I could put the others around these, because the sun still had to be the common focus of all of them. Those placed around had to expand themselves more and more, so that there should remain a space in between for the intersection of those ellipses which lay in other planes.

I compared the two places where ellipses of planets intersect the plane of the elliptic and I clearly saw that it would be more advantageous to remove one intersection much farther from the sun than the other. Then I wanted to place the ellipses in such a way that both intersections were equally great [distant], and then both came close to the earth where the space is certainly narrow and ought to be spared. In the other manner only one intersection came closer to the sun, and to the other intersection I could bring in one more comet. The system became therefore doubly rich in celestial globes. For in the case of the more distant intersections there always remained still enough space. Most of these intersections became pushed back to the superior planets[3] [83] where I found space for many new comets, because one could lay new orbits far out up to the [nearest] fixed stars.

I have not investigated the intersections which the ellipses of comets make with one another. Since these are widening from their perihelia up to their minor axes and [since] this axis lies probably in all cases outside Saturn, a mutual evading of comets seemed to me much more possible because I could see both halves of each ellipse away from the sun as rather drawn out, and therefore I could compare them with the spokes [mentioned] in my previous letters.

You can see from this, Sir, how far I have followed you. I now completely realize that in your system the intersections of ellipses in the zodiac have to be rather scattered, as this is taught to us even by our rather incomplete experience [observations] and from which the now better known orbits of comets are indicated to us.

From this you also explain why the ellipses whose perihelia are close to the earth ought to be very oblong, and, conversely, why there are few planets, and why these few are in one plane, and why the superior ones are more distant from one another. All this flows from your proposition that the solar system should be empty as little as possible. But why [84] don't you also turn now these conclusions around? Do you find the sequence [of your reasoning] strong enough, when you now want to have the correctness of the basic proposition firmly established through so many agreements with experience? I, however, see your method now proven. You derive the basic proposition from the purposes of creation, and what still can remain doubtful in the way of exception about their generality, this you now supplement from experience. The proof from experience considered in itself would suffice if we had a complete record of all comets, or if Halley's Table had been extended to all of them. But since we hardly have the beginning of it, you rely on teleological reasons to supplement what is lacking and to tell in advance what the future findings will teach. From teleology[4] you derive the purposes, and from experience you find that the means for such purposes are available. In such a way the purposes give you the occasion to search for the means, and the means lead you to find the purposes and to judge their universality. You move both closer together and your demonstration thereby becomes complete.[5] So I wished in my last letters to connect in closer manner the abstract reasons of cosmology and the constitution of the real world, because I hoped thereby to go incomparably farther with the conclusions.

[85] It pleased me that my calculation about the heat of the comet of 1680[6] gave you, Sir, the occasion to think about the habitability of this celestial body in a more definite manner, if you had not thought of it long before. Your considerations in that matter were always new to me and very pleasing. You know how many difficulties have been found in that connection. In all appearance those difficulties arose because one stretched the comparison between the earth and the celestial bodies too far. Thus Columbus found after a long sailing a new world and brought the news back that it was inhabited. One could ask him for good reason whether there were humans there. The land was [after all] on the surface of the earth and therefore habitable for men. When, however, an astronomer discovered on the moon, Venus, and other planets mountains, seas, atmosphere,[7] etc., and concluded that inhabitants had therefore to be there, then this conclusion really followed, but one went too far when one also wanted to make humans out of them. One could just as well look for humans in the depth of the sea. We are in general too accustomed to determine everything individually, and the general concept which we should form about the inhabitants of the world is still much too narrow, because we have not seen varieties other than those we find around us on the earth. It is true, their number almost reaches into the infinite; should it, however, exhaust [86] the infinite wisdom of the Creator? The earth is very far from being the whole world-edifice. How difficult it is for us to imagine a thinking being, an intelligent creature, without also giving him two hands, two feet, a head, and other parts resembling a man! If we go far [enough], we add some wings because we feel that we lack the ability to fly. We would even add fins, if swimming were impossible to man and if death were not the result of drowning!

Meanwhile, I must say that to me light seems to spread much too universally across the world that it should serve the earthlings alone. I will not thereby say that all inhabitants of all worlds must have eyes like ours. There can be more ways by which images of visible things present themselves to the soul of thinking creatures, much accustomed as we are to consider necessary the refraction of light and the image on the retina.

The nature of light, its effects, the partnership between soul and body either are not known to us as yet, or are known only insofar as we receive the impressions which the light makes on us, and we seek to fill in the rest through conclusions to which we have as yet no other bases than the ones which we postulate through seeing. At any rate, it still may be that many among the inhabitants of other [87] celestial bodies have eyes, which are only as similar to ours as is permitted by the constitution of the bodies of which they consist.[8]

Another effect, which seems to be connected with the light of the sun and therefore also with that from other suns, is heat and its changes. I know all too well that one considers in that connection still many questions unsolved. Are, for instance, light and heat necessarily connected, or is the latter made only more effective through the motion of light? Does the earth have a basic heat which is native to it, and does the action of the sun contribute only to its yearly changes and to the replacing of that heat which constantly rises and escapes through the air from the land and sea surface? In this case should comets, that come close to the earth and move again away from it for many years, have a similar degree of basic heat, so that the change which arises in each through its orbiting around the sun takes on thereby a less noticeable proportion? The swelling up of the atmosphere of these celestial bodies can still be explained in no other way than through the heat which certainly must notably increase when the comet comes close to the sun. I find it very likely that this atmosphere serves as shield for the comets. It is almost as visible to us as the body of these,[9] and the light must therefore [88] be blocked to a greater extent. The comet, viewed from the earth, has no phases like Venus and Mercury. Its side away from the sun appears to be as bright as the one upon which sunlight falls perpendicularly. On this side the inhabitants of the comet have a real day, and on the other a twilight which does not yield in brightness to daylight. In the same way we have the day[light] of the morning when the sun is half a degree below and above the horizon. Thus have the comet's inhabitants their daylight, but an even brighter one in the sun's vicinity, and this is what we saw about the comet of 1744. The diameter of its atmosphere was nine to ten times greater than that of its body.[10] This atmosphere could therefore considerably weaken the light of the sun and had to scatter it all over the body of the comet. Through this scattering the sunlight becomes many times weaker even when I assume that it all would be refracted to the surface of the comet, which is far from being the case. The scattering of light in the atmosphere may at most serve for warming it up and for expanding it more, but the heat flies upward and seeks the colder regions. It must therefore [move] away from the comet and from the sun, and in its strong flow it presumably tears away a part of the atmosphere, which then constitutes the tail of the comet. This flow is unusually speedy, so that even the length of the tail undergoes a considerable change and the comet loses thereby [89] at each of its returns to the sun something of its atmosphere, small as it can be.

Although I do not believe that the comet will thereby become much poorer regarding its material, still I wanted to find the means for its replenishment. I do not think that the solar atmosphere consists of tails left behind by comets.[11] The material of this tail will remain where it may, but I still would not let it be lost to the solar system. It still remains always heavy toward the sun and may find somewhere an equilibrium. It may be that part of such matter is distributed throughout the ether through which the comets journey and acquire new supply especially when refreezing [in far out space]. This possibility is taught us by experience on our earth. The change in vapors follows much more the variations than the actual degrees of heat and cold. Similarly, the tail of a comet will become particularly longer when the comet falls directly toward the sun. But when it travels through its perihelion, its distance from the sun does not change considerably, [and therefore] the tail becomes shorter and divides itself into branches, and the comet would come to its steady condition if it were not to move again from its perihelion point farther away from [90] the sun. But even through this moving away, the tail must become shorter because the heat slackens and dwindles. Were the comet to acquire new material for its atmosphere in its moving away from the sun, then I would assume in addition to light and heat another substance which would be common to celestial bodies around our sun. Still I do not have to go that far, because very little can be determined in this respect. I certainly see these bodies as very manifold, and I set the greater their differences, the more their orbits are different.[12] I deem it very difficult to explain how a similarity obtains among all of them if the investigation is extended to each individual substance. The universal, as we know, may consist in the laws of motion and in the distribution of light and heat. But from this nothing much can yet be derived about the condition of the inhabitants. I, however, always believe that those on comets must have a strong mettle in view of all changes of heat and cold; and here, I give you, Sir, all my possible agreement that we would be too delicate for that. The pleasant heat of the sun had to be for us more steady, and I compare the earth, our habitat, with [91] the noble friendship that bans all unsteadiness. Such is also the friendship which you have granted to me. How much I also wished to fulfill the other comparison, so that I would again come to you and may constantly stay with you! I beg you to replace with your precious letters the absence of your instructive conversations. I expect those letters with the greatest desire and remain, Sir, etc.

EIGHTH LETTER

I must now also ask you, Sir, whether I did not have good reasons to think of the *sapienti pauca?* Could I very well have after so many proofs any doubt about your readiness to grasp everything with ease? Your whole letter perfectly removed that doubt. How easy it is for you to make even incomparably more difficult notions representable to the eyes through vivid images! How neatly and completely do you express thereby my fragmentary conclusions! I know that all this costs you no effort and would regret my obscure brevity should I think it has aggravated you. Still I must tell you that the various conclusions formulated so briefly pleased me only now and then, and that I further probed them to show that you are in agreement with experience. You will notice that I still would like to submit several of those conclusions to further scrutiny. But the explanation which you gave me about them puts the matter in full clarity. Just as clearly you formulate the manner in which I thought to arrange my proofs. I readily admit that one can call in doubt the generality in everything which I derive from the purposes of creation, if one demands proofs of geometrical exactness. One will [however] grant me the possibility and finally even agree [93] that the conclusions from my system are not contrary to experience. This is also all that I can at most demand. In the purposes which I lay down one finds no exceptions as far as experience goes. Should I, however, extend them further than our experience really goes, one can still certainly object whether it is not exactly there that the exceptions belong. What can I do then? He who wants to see everything before he believes will always stick to this question, and it will always take centuries until the observations, to which I can refer, will become more complete. I could not very well go farther than merely [to note] that these observations can be foreseen.

Here one comes to the difference between insight and habit of thought. Thus, for instance, he who not once looks at the planets as inhabited will admit inhabitants on the comets just as little and even less so. He will always want to see them first, but our telescopes are far from being so powerful.[1] Still you, Sir, already know, that I do not need to be detained by objections of this kind, and among perceptive people the habitability of celestial bodies will be granted without many exceptions. But he who will grant the reasons in this respect will consider me at most as bold for having so much increased the number of celestial bodies in our solar system, and [94] for having set their number as high as this is possible at all. Here the chief question hinges on whether one can extend the reasons given about the planets so far and derive conclusions from them, although we have hardly seen but a most minute part of the celestial bodies of our solar system. This question cannot be resolved with one single conclusion but only in a piecemeal manner, and it always remains to be seen how far its solution approaches completeness. The different pieces of this proof I have already submitted in my previous letters, and I can count with them all those which you took the trouble to investigate and attach them to these.

Of these single proofs one will grant me a few in their full strength. One will grant, for instance, that in fact there are many more comets than planets in the sky. In Halley's Table there are already 21. The comets of 1742 and 1744, whose orbits will eventually be calculated, are not among them.[2] If one further assumes that the comet of 1759 has the shortest period, then it is always 75 years.[3] From this it follows that all comets seen since 1682 were different from one another, because none visited us twice within that time. Most of them need, however, a few hundred years and therefore one should infer very little about those that have been seen since the renewal of sciences concerning their twofold return, and even less concerning their repeated appearance. [95] If in addition one thinks how convenient all circumstances ought to be that a comet might become visible, and how more convenient still if those that are visible to us only for a few days are also to be discovered, then one certainly admits the existence of many hundreds that can hardly be seen from the earth, when many because of daylight, many others because of bad weather, and just as many are not spotted because one is not looking for them. All these must come fairly close to the sun, and I cannot see with what likelihood one would state that no comet stays farther out than Mars, for instance, when the whole sphere [of influence] of this planet can be considered almost nil in comparison with the sun's sphere of influence.

One will just as well concede to me that elongated ellipses which comets describe are more suited to have their number increased, and that thereby the variety and greater diversity in our solar system will become far greater. One will admit that as soon as the number of these bodies, through the safeguarding of the law of gravity and of their orbits, ought to be in all cases the greatest, the very oblong ellipses become necessary and ought to occur in greater quantities. Why do they, however, actually exist? It is very conceivable that the law of gravity, the habitability of the solar system, the variety and periodic changes in the celestial bodies stand [96] in such a mutual connection that taken together they constitute a maximum.[4] One is not able yet to show why the Newtonian law of gravitation rather than any other law was brought about in the world. About various other laws Newton investigated the consequences and their unsuitability.[5] If, however, a rigorous demonstration is required, I am much inclined to think that the habitability of the solar system and the

96

variety of orbits will be counted among the basic reasons. A spiral orbit to which I have recently given some thought would take up as much space as yet uncounted ellipses. However, when a celestial body enters into an elliptical orbit, all the space around it remains free and still innumerable other ellipses can be added in the same and other planes. Each spiral orbit would retain a whole plane, and if its circuits would be close to one another, as this necessarily has to be so near the sun, then not once could there remain room for intersecting with other spirals. Each celestial body that had to move in such orbits should be specially set for each [specific] distance from the sun, in spite of the fact that it could remain at such a distance only for a moment.

On the contrary, how much more suited is the Newtonian law of gravity for [securing] habitability and variety in each solar system! How much more harmonious is the return of orderly disposition in each periodic [97] orbit! How many varieties similar to one another [are there] among the celestial bodies that stay around the sun! It is always ellipses, well-defined periods, succession of seasons of such duration, which bring about on each celestial body the changes ascribed to them.

Should, however, even higher viewpoints be reached, and should the distance from sun to sun be sized up, and should the world-edifice, the system of fixed stars in its entirety, the ground plan of the world and its arrangement be considered, then this law of gravity still will do and will give the shortest way to it. If you transform the ellipses into hyperbolas, the celestial body which moves along it does not stay with any sun. It hardly bends its path to turn toward other suns. For this change of direction it needs but the shortest time, because it moves at the highest speed. Then it approaches its asymptote and enters along a straight line into the [gravitational] area of another sun where its velocity again increases only to move soon again toward new [solar] systems. Could one here think of a more convenient route than hyperbolas and of a law of gravity more capable in every respect than the one which is demanded by conic sections, the simplest of all curved lines of which one half is periodic,[6] the other, however, gives ever new varieties, and with both of which all possible variations can occur?

[98] I believe that one can here conclude that these changes in the world do not remain merely possible. The [principle of] variety has to say much too much in the doctrine about perfection than to permit me to omit to a large extent those changes from the world-edifice. This would happen if I limited the number of comets to a few hundred in spite of the fact that millions of them are possible without disturbing one another. About comets whose perihelia are equally distant from the sun some can stay in the outer regions much longer than others, because the longer axis of the ellipse does not depend on the focal length. Again, different comets

can have similar periods but very different seasons, because the longer axis alone determines the time of their revolution but the focus can be at any point of that axis.[7] How many changes and varieties are here, too, possible! Should I exclude by far their greatest portion? For this I have no reason; on the contrary, the greatest perfection also demands the greatest fullness. It must have all that it can have. How much can the distance of the real from the possible still amount to and what else will be lacking to my system than the complete register of all comets, if one does not consider these conclusions as adequate prior to experience?

It is true that we have every reason [99] to be cautious in natural science, and experimental evidence makes itself needed in that science in a special way. One has long since become accustomed to considering only as probable such conclusions as are not based in their whole extent on experience, and most of them are derived only from general considerations. So one held in older times the [spherical] figure of the earth, then its revolution, and afterwards the inhabitants of planets to have been no more than probably demonstrated, and as to the latter there still lacks the autopsy[8] for all those who do not believe without seeing and for whom general proofs are not enlightening. I can therefore present my system only as probable, but I still have the advantage that one will now make more effort to let no comet go by unobserved. The register will be more complete, and anyone who finds my conclusions worthy of approval will admit that this register will corroborate its correctness only after many centuries. Each comet which is different from the previous one belongs to the balance which in future times will obviously tilt.[9]

With all this, such a register will still remain incomplete in two respects. First, there will be absent from it all comets that do not come close enough to us to be seen, and therefore in my view their greatest portion, because I find no reason why there should be no comets outside our [100] sphere of vision. Those which one spots occasionally with a telescope may in a large part belong to this class. Therefore when I state that there are also comets with hyperbolic orbit, [I also state] that we see all these only once. One can through careful computation determine the orbit and on that basis distinguish these comets from those in an elliptical orbit only when the difference is in fact considerable.[10] I think, however, that such comets do not come so close to our sun, or that this happens only very rarely, [and] since the ellipses do not reach so far as to come close to the sphere of influence of another sun, it seems that the space outside them caters mainly to the hyperbolas. Still nothing definite can be said on this matter and the former case always remains a possibility. Should, however, such hyperbolic orbits also pass through closer to the sun, then I would assume such an arrangement in the world-edifice and set their approach at such a time that they could retrace their path undisturbed, because I wanted to preserve for the inhabitants of each celestial body the vantage point set for them for the contemplation of the world-edifice. I would always permit smaller displacements,—the major ones are possible considered in themselves,—but I assume in the world no chance event but [only] purpose, order, and the wisest and best arrangement. Here one can consider me too generous, but anyone who admits the contrary may as well investigate to what extent can he assert [the possibility of] new creations.[11] [101] For I would not allow any celestial body which ought to remain forever barren and devastated. I do not at all think that a celestial body can make itself inhabited, or that, should the earth become a comet, the seed for industrious creatures could lie hidden in it and sprout up afterwards. On the contrary, what I have observed in my previous letters about this arrangement of planets and of their satellites serves me as a necessary proof of their permanence. They all orbit from west to east, their orbits are alike very round, and they are as numerous as permitted by the smaller perturbations, [and they are] in one plane as they ought to be if [enough] space is to remain for more comets. This I call a planned layout which has purpose for its basis and not mere chance where all purpose is lacking.[12] I extend this layout without hesitation to the hyperbolic orbits of celestial bodies. Undemonstrated as it may seem, one can confidently assign it to the infinite wisdom of the Creator, and it will very much provide us with material for a wonderment full of awe. Has He not provided for us on the face of the earth in innumerable ways? Should His providence for the dust be vast, but should He be negligent about the whole world-edifice which needs that providence through eternities while we hardly need hours? I must, however, return to the matter which I wanted to prove in part.

I also note that one will grant to me [102] that the solar system will be much more habitable if there are fewer planets and these few move in the same plane. You have, Sir, given to this proof all clarity so that I need not first look for it. These [following] few details still occurred to me in this connection. You know that the sun turns on its axis and that its equator is inclined to the ecliptic under an angle of $7\frac{1}{2}$ degrees. Since the orbits of all planets lie in almost the same plane, and the inclination of the sun's equator is very small with respect to all of them, and physically speaking one can put the sun's equator and the orbits of planets practically in one plane,[13] the smaller displacements in the orbits of planets permit no geometrical symmetry. What chance can bring about this consonance, or do the planets require thereby preference over the comets? Chance events are never the case with me. In my system the planets all had to orbit in the same plane, and should I choose a plane, the sun's equator would be selected for that role. As of yet I perceive no other reason for this than that it would be a start toward the harmony of the system. I do not, however, doubt that there may be other reasons in that connection. The reason which I have in mind I would rank only with the least important ones, because I cannot prove that it is weighty. I must assume that the sun can undergo in single parts [103] of its surface noticeable changes of brightness. A great change in brightness would bring about an equally great one in the heating up of planets. Through the rotation of the sun on its axis this heat is uniformly distributed among the planets because each day other parts of the sun's surface come into their view. This would not happen in so orderly a fashion if the planets orbited in a plane greatly inclined, or simply perpendicular, to the sun's equator. In the case of comets this matters but little because they are certainly accustomed to greater changes of heat. If the heat of the sun, inasmuch as it is different from its light,[14] extended as far as does its atmosphere in the plane of its equator, or if this atmosphere served merely for the spreading of its heat, the planets would certainly profit from it in a [more] uniform manner than if their orbits were more strongly inclined to the sun's equator.

About the inhabitants of celestial bodies[15] I was not much concerned, because I saw it well that our concepts were not rich enough to describe them each according to their circumstances and according to the conditions of their habitats. The remarks which you, Sir, have made in that connection pleased me very much. I see with you the usefulness of light as a very universal one, and would the ability to absorb it be granted to the inhabitants of each celestial body, it might be transplanted into their souls either through the eyes or through some other means. Here again I state [104] nothing definitively, because we cannot form a notion of any other means.

It may be that in addition to light there are still several other substances common to many or all celestial bodies and different only in their modifications. We still do not know in how many ways those substances can change their shape once they are mixed with others. In general, I can state that those are common which spread out through the solar system or through cosmic space. The vapory layer of comets may together possess something similar, especially when in returning from the sun they start on their departure again. For with respect to celestial bodies I am together with you, Sir, mindful of the preservation of the whole. The water on our earth dissolves into vapor, the vapor into pure air, and this perhaps into a more subtle substance. The finest solutions can then take the return path and change into water. So it may also be with comets. Should there be on comets liquid substances similar to our water, then in order to maintain them as liquid I would give the comets considerable ground heat, and the dissolving of that liquid into vapor should serve the comets as a shield against the heat of the sun. Meanwhile, we can make no certain judgment about the condition of those bodies. The vapory layer of comets always remains transparent because we can see through it, be it even 8000 miles high, the very body of the comet.[16] This must still therefore be strongly illuminated when it comes close to the sun, [105] although I similarly think that the heat originating there always moves upward through the atmosphere because the light is still very much scattered there.

You see from this, Sir, that I do not really dare to adduce in my system something individual, but only a ground plan of its arrangement which would come closest to the true one. The difference in the case of each celestial body must in general be great, because the Creator spreads out through the world-edifice the choicest features from the riches of His omniscience. His goodness and wisdom must necessarily achieve in each celestial body the sum of what is suited and useful to its orbit in order to equip it in the fullest manner. We see the proofs of this on our earth, and just as well but with infinitely many variations shall those proofs become recognizable and worthy of worship to thinking beings on each celestial body which they inhabit. I revere in this especially also the eternal goodness that brings together harmoniously disposed souls, as ours are, in time and place. May your friendly wish be fulfilled that so small a difference in place may never separate us! I hope and look longingly forward to the moment that will permit full outburst to my joy which I feel to a lesser degree as long as I must present myself from a distance, Sir, etc.

[106]

NINTH LETTER

Do you, Sir, then never rest when there is occasion to make your world-system acceptable to the unbelievers? In your previous letters you have investigated to what extent the general purposes of creation, compared with our experiences, can bring the proofs to completeness. And now you weigh the various viewpoints from which the arrangement and habitability of the world will be considered. I agree perfectly with you that those who still doubt that planets have inhabitants, or simply deny this, are still far behind and too narrowly confined in their thinking, because they know apart from their eyes no other means to necessitate approval and therefore want to hear nothing of general proofs and of moral certitude. It is not even necessary that everybody should understand everything, and there can be among earthlings a difference on a small scale similar to that which exists among inhabitants of different

101

celestial bodies. Those, who follow hyperbolic orbits and wander from sun to sun, extend their consideration to the whole, while we mostly stay with the earth. Our horizon is in general narrowly confined. But does it not have innumerably many smaller degrees? Most people know only a few villages, their birthplaces; [107] others may have seen greater stretches of land. A few swing in keener reflections above the air, and even fewer penetrate through the depth of the firmament. You, Sir, pushed through it and now you seek how to make your discoveries comprehensible to each narrow view and how to widen each horizon to the point where it will come closer to your considerations.

You may, Sir, be reassured on my part that willingness is not lacking in me to follow you and extend my horizon through all worlds, if you proceed with your conclusions beyond the fixed stars visible to us, nay beyond the wonderful light which Derham and others have discovered in Orion.[1] My first letter should serve you as a proof that I have already busied myself a great deal in using the yardstick [appropriate] for such distances and to let myself be carried on each light ray through all worlds. I have a special pleasure in seeing the connection which you in all appearances further extend to the whole. It is not enough to you that the world be made into an interdependent whole through the law of gravity. You also tie together each solar system with one another in that [respect], too, that none of them should be separated from another in such a way that all its globes[2] should remain connected with it. This you grant at most only for one half [of those globes], or even for a much smaller part, [108] and you let the other part, destined to much higher considerations, move from sun to sun.[3]

How much more the world becomes inhabited than one would have thought only a short time ago![4] We find worlds in each speck of dust, in each droplet, and soon the specks of dust will not be as numerous as the celestial bodies in the firmament. It is unquestionable, Sir, that he who grants inhabitants on the planets has no more new steps to make to surround them with as many celestial bodies as you wish. For he will grant that the world should not remain barren and that there ought to be in it more live than dead masses. The Creator, the eternal source of all life, is much too efficient not to imprint life, forces, and activity on each speck of dust. How should one then consider your enterprise mistaken when you, Sir, do nothing else than show that, if one is to form a correct notion of the world, one should set as a basis God's intention in its true extent to make the whole world inhabited and leave no part, no side [section] of it out of consideration! The only thing that one can object to the universality of this viewpoint is the concern that some higher and to us unknown reasons may prevent letting so many celestial bodies wander around each sun. He who is concerned in that respect will certainly wait for the complete register of all comets, and that will take [109] many centuries, and for all that one shall never draw it up in a complete form.

I have made for myself a quick analysis of this register. From 1500 until 1600 one observed over 40 comets of which it seems none reappeared for a second time during these hundred years, because I still consider the one of 1759 as the speediest and I assign to the rest more than 100 years but to most of them many centuries for the completion of their orbits. According to that calculation I shall have in 400 years 4 times 40, or 160 comets. I set at 60 [the number of] those that among these return twice or several more times, so there still remain 100. For these 100 I must, however, take at least 300, because so many hindrances stand in the way of the visibility of a comet that I can assume that we see of those which can be seen hardly a third. Thus I had at least 300 comets that descend to our sphere of visibility. This sphere is, however, as you, Sir, remark in a previous letter, 40 times smaller than the sphere of Saturn, [and] consequently I had to take these 300 comets another 40 times to account for all those that come within the sphere of Saturn. The calculation gives 12,000 comets. Although this number is far from reaching 5 million, it is nevertheless a legion and the number of planets becomes negligible by comparison. Meanwhile, I do not really believe that the register of comets visible to us [110] amounts only to 300. On the contrary, I think that it stretches to several 1000. Halley's Table shows such a variety in the planes of the orbits of comets as to make us suspect very considerable gaps there. Thus, for instance, the comets of 1672 and 1698 had in their perihelia almost the same distance from the sun, but they were in all other respects completely different from one another. And exactly this holds also of the comets that one has seen in the years 1532 and 1596. One can also compare all the possible differences, and the gaps that remain in Halley's Table will become obvious. These 12,000 comets all still come closer to the sun than Saturn. Nothing prevents, however, that there should not be such comets which stay out 10 times farther, and the habitability of the world demands that they should be all the more densely spaced, the closer they are to the sun. In such a way I had to multiply by 100 these 12,000 and I will come up with 1,200,000 comets[5] even when the register of those that can be seen by us does not stretch beyond 300.

Comets that come close to the sun still leave so much space outside Saturn that I would not hesitate at all to place the greatest comets in that outer region and add satellites to them. For I do not believe that a comet with satellites would ever come into our view. The [111] closer they come to the sun, the narrower will be the space and the more sparingly ought it be used. A comet that has satellites around it has a very large sphere of influence, but it still must leave room for other comets. This is, however, much more possible if the comet stays farther away from the sun. I think, as you do, Sir, that order, diversity, habitability, and the law of gravity determine one another in each solar system, and this is necessarily the reason why you consider only the general, because the individual things must be discovered through experimentation.[6]

However, I grant you, Sir, that our solar system, as far as presently known to us, already shows enough worthy of wonderment. For I refer here also to that [point] the reason of which we still do not perceive completely. The planets seem to have something special to themselves. Their orbits are in a plane, as you remark, in the plane of the sun's equator, and all, even their satellites, orbit from west to east and in ellipses which differ very little from circles. The comets, however, orbit in sundry ways across one another and have all possible directions in their orbits. This fourfold order and harmony in the planets necessarily excludes all chance. At no other time have I perceived more fully that comets are not [112] immature planets and that they do not turn into such through a displacement in their orbit. It is 65,536 times more likely that of the 16 comets, that should have turned into planets and satellites, at least one should move from east to west.[7] The impossibility would still become infinitely greater, if I were to calculate how unlikely it would be that all 16 comets have suffered perturbations in such a way that finally they started moving in one plane and, in addition, exactly in the plane of the sun's equator. All chance necessarily vanishes here and we should consider the ordering of planets as the result of a law that has its neat reasons and purposes, hidden as these may still be to us. That the orbits of planets are hardly different from a circle can finally be grasped from postulating that the variety of inhabitants in the solar system demands also such inhabitants, who need moderate and always equally great heat and light. That they all move in one plane, you have, Sir, derived from the consideration that thereby the habitability of the solar system becomes enhanced. That they all move from west to east is a fact of which in all appearance no reason can as yet easily be given. The law of gravity does not seem here to suffice, because there are comets which have a quite contrary direction in their motion as, for instance, the comet of 1759. One also used this reason [113] to overthrow the Cartesian vortices.[8] Should this movement common to all planets have a connection with their being in the plane of the sun's equator, or does another law of the world-edifice lie hidden here [in the fact] that the planets both in their orbits and around their axes must turn in that very same direction in which the sun turns on its axis?[9] Something so universal presupposes a universal reason and a universal law, and this law seems to be necessary only with the planets but not with the comets, because these do not all move from west to east. For I do not think that the skillful evading of one another by celestial bodies alone demands such a harmony in the motion of planets, or that now some comets move from east to west only because they were diverted by Jupiter or by one distant great comet, so that they have no longer that direction. Still this is not impossible. The inhabitants of comets seem to be unaffected by heat and cold, and longer or shorter winter should not mean so much to them as to us. But [shifts like] these would be only exceptions which ought to occur in very small numbers. I would therefore postpone decision on this point until the register of comets becomes more complete. Then one would see from it whether in fact only a few comets go from east to west,[10] or whether there are almost as many of those [114] that have an opposite direction. Meanwhile, so great displacements seem to me very unlikely only because I would prefer to let each celestial body remain what it actually is. I could, of course, soon again worry whether the earth might not also be subject to such a displacement that would threaten it with disruption. I do not, however, let this apply, because the earth and the rest of the planets never were comets and much less can they turn into comets, because they are not fit for that at all. On this point I am now completely reassured and I shall invariably abide by the reasons which you have given me in your previous letters.

How excellent the Newtonian law of gravity becomes through the considerations which you, Sir, have presented on it, and how precisely you connect them with other purposes of creation, in terms of which you assess the arrangement of the whole solar system and similarly predict what only posterity will carry into the realm of experience! I wished I could determine the maximum which you have mentioned in that connection.[11] You connect the law of gravity with the habitability and diversity of celestial bodies in such a way that these general purposes of creation certainly seem to require that law, which anyhow is the simplest and facilitates so much for us the determination of the orbit of each celestial body. How admirable become thereby the arrangement [115] and order in the whole solar system, and what a manifold diversity and variety are spread through that law across the whole world! Thus light is shed on the choicest of all orderings, which should necessarily appear to us as disorder if we, unaware of that law, could survey the solar system at a glance in such a way that each planet and comet would come into our view. In fact, it would present itself to us in no other way than as an entangled and disorderly scattered heap of globes which were unrelated with respect to size. We would find no reason why their distances from one another are so unsimilar and in all appearance so clumsy. I cannot sufficiently exert my imagination to present to myself the position of these celestial bodies, as it is, for instance, at the present moment, and to track down each comet in its orbit. This much, however, I think that it would appear to me as if they all had gone astray and many of them were at one place and almost none at another.

But I take the law of gravity and the resultant order in the course of celestial bodies that are around our sun, and this erratic picture at once vanishes. For this law teaches me that I must not consider the celestial bodies in themselves but also their orbit and the law of their motion, and then I certainly [116] find the exquisite harmony which you, Sir, point out so worthily in your letters. How clear it becomes from all this the great tenet: the disorder in the world is only apparent, and where it appears to be the greatest, there the true order is even more excellent though only more hidden to us! We shall feel, when we consider the members of the solar system only with respect to space and position, as if one were in a well-ordered library where one only considered the place to find each book simultaneously. This order would be much too simple in the world. Here space and time must be connected with one another,[12] and the order ought to be extended to both. How perfectly harmoniously this occurs in the motion of celestial bodies around the sun, which, from the viewpoint of space, do not seem to have any order!

Do you think, Sir, that these considerations may not be extended to the fixed stars as well?[13] I viewed the starry sky in this respect only yesterday evening, because I have never yet been able to find a definite symmetry in its apparent shape. I looked for it again but in vain. I found the stars of the first, second, and subsequent magnitudes so unevenly distributed, since many of the brightest stars were close together, whereas in other areas of the sky there were great empty spaces and in these there were hardly a few stars of the sixth magnitude. So disordered, I thought, would the solar [117] system appear to us if we could see all its planets and comets at once. But we know that in fact there is an exceptional order in it regardless of how differently it may appear to us. Along what laws might the All-wise have sown these eternal lights through the immeasurable depths of the firmament, through the magnificent vestibules of His dwelling place? Should I here, too, bind time and space together, and consider the fixed stars no longer fixed but as suns that wander in majestic circles and need aeons to complete a step, or to travel one degree of their orbit?

Mostly I paused at the Milky Way. This luminous arch, which stretches all around the firmament and decorates the world-edifice like a ring studded with gems, roused in me astonishment and wonderment. Just as on earth we see the rainbow produce the image of the sun in innumerable droplets, so the great Creator seems to have extended around the heaven the droplets of light in which He dwells. How densely, how immensely densely are they here together and how desolate does the sky appear outside this luminous streak! Here I became confused and I must tell you, Sir, that soon I began to doubt again the universality of your basic tenet that the world is as habitable as possible and [118] ought to be inhabited. If there could be so unthinkably many suns in such a narrow streak, how could it fail to happen that the whole sky is not covered just as densely with suns? You see, it now goes with me entirely as with the philosophers whom I castigated in my first letter.[14] I was brought unawares in their footsteps and soon I started heaping questions and doubts upon one another. Still I hope that my doubts are not so terrible and, if only the comets leave us in peace, I am not concerned about a fixed star ever visiting us. It would then declare war on the whole solar system, but I would never allow this.

What do you think of this and to what extent do you, Sir, feel confident in justifying your basic tenet about the habitability of the world-edifice and also in showing that it is connected with the order and arrangement of the whole and with the law of gravity? Yet how wonderful must things look in the Milky Way! Do you believe that the stars in this luminous streak are in fact more densely together, or do they merely lie in endless long rows behind one another? Whether the former or the latter is the case, I come always back to my previous question, why is it so only in that single streak? Innumerable as are the stars outside [119] the Milky Way, compared with those in this circle they are hardly like a droplet against the ocean. Our best telescopes are not able to present them densely enough, and everything that can be distinguished there simply amounts to the bigger stars that are in the Milky Way and to the shape of that streak as it appears to our eyes. The outer contour of it is very irregular, its width is very uneven, and in some places hardly 3 degrees but in others 25 to 30 degrees. It appears as if put together of pieces, and some pieces seem to be so distant sideways from the rest as if it were split up or torn apart.

Can you, Sir, find harmony and order in this apparent irregularity? And how would you connect here time and space with one another, so that all this would be governed by one universal law? For nothing seems clearer to me than that here, too, such a law and a wise arrangement should be present, and I can never imagine that, when each part has the most perfect order, this order should be missing in the whole. The world-edifice is a whole and ought therefore to be necessarily interconnected through universal laws. I am writing for your thoughts on it with more expectation than for anything that I still might learn about our solar [120] system. This will clear itself shortly and will bring your predictions to fulfillment. But in which future millenia shall those of your posterity live who discover a change in the position of the fixed stars and deduce from it the laws of the whole firmament? You can, Sir, spell out your considerations all the more boldly because I am not concerned about the judgment of so remote a posterity. I know that you always fall back on the most likely and this is for our times already something very significant in such far-reaching matter. I will assume with complete recognition that I am indebted to you for so many instances of kindness, and will receive them with always greater animation as long as I live or can sign myself, Sir, etc.

TENTH LETTER

You give me, Sir, a much desired opportunity to let my imagination soar freely and you are even willing to justify this, because the experimental evidence on the basis of which one could oppose me is still very far away. But you know that I love too much that probability which is still admissible with more rigorous investigation, and I would leave it to the poets to make an astronomical novel from my reflections on the world-edifice. Many would not simply take them for anything better.[1] I will therefore explore how far these conclusions can reach and will supplement the rest with speculations, because the full clarification of the questions submitted by you will be reserved in all appearance to times to come. Meanwhile, it is a true pleasure for me that you, Sir, now also begin to set no more limits to your noble thirst for knowledge, but extend it through the whole world-edifice and look for eternal laws and for wise order even there where mere confusion seems to rule. Continue posing questions to me, because your investigation leads me to further reflections. But help me also to broaden my considerations and to make them more complete by whatever still may occur to you in the matter.

[122] To start with, I remove the fixed stars from their positions and make them wander in specific orbits much as the case is with planets and comets. You have, Sir, already given me in a previous letter a reason for doing so, because you extend the law of gravity through the whole world-edifice, and assert that the world and all its parts become thereby most closely interconnected with one another and turn into a coherent whole. This law gives, however, nothing else but a centripetal force, and were it the only force, all celestial bodies should gradually fall together into one lump. In 64 days the earth would already be in the sun. In this way the preservation of these bodies would be badly taken care of, and it is therefore necessary that the law of gravity be so limited by another that those bodies may endure. This [other law] is the motion and the centrifugal force arising from it. These two forces hold one another in a specific balance, and the order among the celestial bodies will thereby become steady. Should now the fixed stars gravitate either toward one another or toward a common center, they would not remain at rest but would of necessity move in circles because of a two-fold force.

The other reason you have submitted, Sir, in your last letter. You find in the position of fixed stars no symmetry with respect to [123] space and position [time], and from this you conclude that the order must connect space and time, and therefore the order is more perfect because it is more coherent. You illustrate this with the solar system which should certainly appear to us as confused, if we could see all the celestial bodies that go around the sun and if we were to stay with that mere look. How would you then want the fixed stars to have a co-ordination among themselves with respect to time if there was no relation, no universal law between time and space? This, however, necessarily presupposes some motion, and I conclude with you that the fixed stars are far from being as motionless as one makes them out to be, but that they are suns which wander in prescribed orbits.

This motion of the fixed stars must necessarily produce a change in their apparent separation from one another in the course of time. I cannot say that since Hipparchus, who composed the first catalogue of fixed stars,[2] anyone has observed such a change.[3] So much is certain that these ancient observations do not agree with the more recent ones, and one has looked for the reason of this only in the less precise instruments, and later also in the fact that the refraction of light, the aberration of light, and the nutation of the earth were unknown to the ancients. If one could improve these errors, or at least investigate whether the difference between the ancient and recent observations [124] is greater than the one coming from these errors, then the question would be settled. It is clear that one had to compare above all the positions of the greater fixed stars, because these are more likely the closest to our sun. Since I do not now have the time to undertake this investigation, I stay with the more general reasons which I have previously adduced. These seem to me satisfactory to show that the fixed stars must change their positions with respect to one another, and much more cannot be arrived at from the investigation just mentioned. Here, too, as in the case of comets, I stay with the general arrangement.

This motion of fixed stars can be considered in a twofold manner. Since there, too, the law of gravity obtains, they move, [as] they belong so much together, around their common center of gravity just as this takes place also in our solar system. The question now merely becomes whether one leaves this center completely void, or whether one should put there a body very large compared with the fixed stars which revolve around it, which is exactly what the sun is in comparison with the celestial bodies that orbit around it. Were this center quite empty, the motion of fixed stars ought to be very slow, because they would have with respect to that center no other centripetal force than the one arising from the fact that they gravitate toward one another. This gravity would, however, be [125] very small because of the great distance, and therefore the centrifugal force and consequently also the [orbital] velocity would not be great. The planets and comets of our system would therefore move around their common center of gravity even if the sun were not there. But their velocity would be incomparably slower because [otherwise] they would soon be scattered.

Should I, however, put in the common center of the fixed stars, which together constitute a system, a body toward which they all gravitate, then I must give that body an enormous size[4] and I must enhance its mass so much that even the most distant stars of the system would have considerable gravity toward it, because this is always proportional to the mass. Were I to write here a novel, I would then state that this body has either no proper light or only a very weak one.[5] I would so arrange the world that the smaller dark bodies, like, for instance, the planets, would move around bright suns, but these again would orbit around dark bodies. For the suns would need no other light because they themselves possess so great a brightness, while the dark body can still sufficiently be illuminated by the suns which move closest around it. But for such an arrangement I can give no reason other than the mere possibility. You know, however, Sir, that [126] possibility is considered satisfactory only in the poetical but not in the philosophical world. I leave it therefore undecided whether a system of fixed stars revolves only around a central point, or whether at that point there is in fact a body of enormous mass toward which the fixed stars gravitate.

Another reflection which may be made about the motion of fixed stars is based on the question whether the motion of a celestial body around its axis is necessarily connected with its motion in its orbit, and whether the mechanical arrangement of the world-edifice makes the one dependent on the other. The relation between both these motions one cannot yet determine from general reasons.[6] Should this, however, become possible, then I would soon be ready with the conclusion: the sun moves around its axis and therefore it also must go into an orbit.[7]

So much is clear that the sun, together with its planets, satellites, and comets, moves around a common center of gravity. This follows from the fact that all bodies gravitate toward one another. But the circle which the sun describes for that reason is necessarily very small, and one cannot conclude therefore that its center should gradually be displaced. This would anyhow occur very slowly. I therefore prefer [127] to derive the motion of fixed stars from the first two reasons which make it more considerable and more necessary.

What you, Sir, write to me about the Milky Way, has already astonished me repeatedly. It seems unquestionable that this band is much behind the other fixed stars which we see outside it, and the bigger fixed stars in that circle ought to be incomparably closer to us because only with the aid of telescopes do we see more distinctly those which specifically form the Milky Way.[8] Since, however, they are so immeasurably distant, it seems certain that they do not yield to our sun in size and brightness, and therefore they must have a considerable distance from one another.

Therefore start out in thought from our sun straight toward a fixed star of first magnitude, from that to a more distant one, from that to those that are farther according to the order [of magnitude], and you will perhaps sooner come to the outermost [star] if you had taken the road outside the Milky Way rather than the one going through it. I set the nearest star, which properly belongs to the Milky Way, many times farther away from us than the farthermost among those that do not belong to it. For since I assume that the stars in that band are as much separated from one another as any of the nearest stars are from our sun, I must [128] therefore necessarily put them in inconceivably long [many] rows behind one another, and thus I conclude that the whole system of fixed stars visible to us is not spherical but flat, roughly like a disk whose diameter is many times longer than its thickness. For I must take here a physical plane which has a certain thickness. In that plane lies the Milky Way and all visible stars outside it; that plane is also the ecliptic[9] where all fixed stars move.

But this is far from enough for me. The Milky Way distinguishes itself clearly from the rest of the sky. When I take together all the other fixed stars, I must completely separate from them the Milky Way and I also must divide that band into innumerable smaller parts. Many of these parts show themselves to us through the fact that they appear separated from the rest. The others cover one another because one lies behind the other. Each of these parts I see as a separate system of fixed stars.[10] We find ourselves in one such system and I count into it all stars that are visible to us and lie outside the Milky Way as well as the larger ones that cover that arch of the sky. I set each such system similar to our solar system in the sense that all fixed stars or suns which belong to it move around a common center, and [129] I would be inclined to believe that all these systems, or the whole Milky Way, have a common center around which they move.

You will see from this, Sir, that I make my conclusions on the basis of analogy. Thus, for instance, the satellites belong to the planets, these to the sun, the sun to its system, and this to the system of the whole Milky Way. Farther our eyes do not reach and I leave it undecided whether the Milky Way visible to us still belongs to uncounted others and forms with these a whole system. Perhaps the light of this so immeasurably distant [super] Milky Way is so weak that we are unable to see it. For the nearest fixed stars still can spread a weak light through our air, as we can see at night when the sky is clear. This so weak a light can unquestionably obscure and make insensible a still weaker light, and I conclude from the fact that the Milky Way is still visible that in that streak there ought to be innumerably many stars. To the naked eye the stars of the seventh and subsequent magnitudes vanish and they become visible to us only when many of them are densely together. We see this in the so-called nebulous stars. The telescopes teach us that they are but a heap of stars which the distance makes too small for us to distinguish with the naked eye. When, however, the light from many of them fuses together, [130] it becomes stronger and therefore perceptible to our eyes. This is the reason why we see the Milky Way.[11]

Such is the appearance of the firmament as I present it to myself, but I readily concede, Sir, that I have not yet sufficiently developed the reasons for it. It will be a pleasure for me when you help to investigate it more adequately. Since here I bind time and space together, I hope you will not be taken aback by the fact that I did not sow the whole sky with stars as thickly as the Milky Way. You have already given your approval when I let the number of comets increase not as the cube but only as the square of the distance of their perihelia, because, as you yourself realized, I had to leave space for movement. Conversely, could I not also draw from this the conclusion: the heavens appear to be empty and desolate outside the Milky Way when compared with this ring, consequently the space there is reserved for movement and the fixed stars and all their systems must have definite orbits? For you will be able to derive from all my previous letters that I do not admit more celestial bodies than orbits are possible; but where there is still room for these I do not hesitate to place there a celestial body and as a result bring the world to completeness to that extent. Motion seems to me to be essential to the perfection of the world, for thereby changes [131] arise and with changes ever new varieties. The motion brings forth these varieties, the laws of transformation are universal and simple, and therefore we have the consonance of the diverse, perfection that is. In such a way cosmology, too, confirms my proposition about the motion of fixed stars.

Sir, you have helped me to extend the Newtonian law of gravity through the whole world and also through each system of fixed stars. Just as far do I extend the circles and ellipses in which the fixed stars move around the center of their system. The closer a star is to that center, the faster it moves in its orbit and the shorter time it takes in moving around. I have already remarked above that I take the stars which are outside the Milky Way together into one system, great as their number may be. It is natural and consonant with the [principle of] analogy that number, space, and time grow together with the system. The earth has only one satellite, Jupiter 4, Saturn 5, the sun many millions of celestial bodies around itself, because its mass is [so much] greater and its heat and light are of more general benefit. Should I put together accordingly a system of suns, millions of them are still far from being sufficient. Their number must be incomparably greater if it should increase [132] in the proportion which the satellites of Saturn have to the number of bodies in the solar system, which I have set at millions in one of my previous letters.

I cannot determine whether our sun is near the center of the system to which it belongs, or whether farther away from that center. This much can be stated that the sun is not near the extreme borders, if I can conclude according to the model described above. For I placed between this system and the rest that lie around it in the plane of the Milky Way a wide space by which the systems are separated from one another. Would our sun be near the border, we would see only in one half of the heavens fixed stars of the first, second, and subsequent magnitudes. This is, however, not so because the starry sky shows us such stars everywhere. Therefore our sun is nearer the center of its system.

On the contrary, it may very well be that this whole system does not lie within the plane which goes through the Milky Way. For were this so, the Milky Way should appear in the celestial sphere as a great circle, and as a result it would be equally distant from the poles of the equator and of the zodiac and would divide the equator and the zodiac into two equal parts. This, however, does not happen. Therefore [133] the Milky Way presents a smaller circle and our system lies outside the plane that goes through the Milky Way. Since the visible form of this streak is not too different from a great circle on the celestial sphere, our system cannot be far removed from the said plane. Perhaps it lies as much outside as the piece which we see protrude sideways from the Milky Way.[12] If it were farther out, the Milky Way would appear much wider. Its mean width amounts, however, only to about ten degrees.[13]

I have not yet pursued these considerations any further and they cannot therefore be presented here in a more elaborate way. I also submit them only as arbitrary and I have perhaps described them unclearly and without a proper order. But you know, Sir, how difficult it is to extend order over a matter to which most pieces are still lacking. You will oblige me very much if you share with me your thoughts on it with your customary lucidity, for I know that you have an outstanding clarity in all your presentations. Your assistance will be to me a new and unforgettable proof of the friendship with which you regaled me and which always renews in me the pleasant impressions with which I am, Sir, etc.

[134]

ELEVENTH LETTER

Now I see in its full extent why you, Sir, have always said that we are still far from thinking in a truly Copernican way. It would not be enough to stir the earth from its rest; rather, not a single body in the whole firmament should remain at rest. The sun may forever be in the center of its system and let the planets and comets wander around itself. The sun is no more than Jupiter and Saturn are in respect to their satellites. But that the sun should be in the center of the whole world-edifice is far from evident; and should the sun once happen to be there, it would soon be removed. In short, resting is banned from the world-edifice because that would make it too uniform. Variety demands changes and these cannot occur without motion. Motion becomes therefore essential, and the general law of gravity will suffice to show that everything is alive and moving. No point of the whole world-edifice remains, even for a moment, in absolute rest. The most complete symmetry must connect time and space, and each dead [motionless] mass will be excluded outright from the world, and [for] without a universal motion the world would be an unused machine, a rundown clock.

[135] So far, Sir, I grant you everything. The only question is to determine the manner of this motion more exactly. You have already sufficiently provided for our solar system, and from the general ideas of its arrangement you have also predicted what experience will increasingly verify as time goes on. I can just as well concede to you that you have rightly inferred from the law of gravity, which extends unquestionably across the whole world and makes it into an interconnected whole, a central motion of the fixed stars, little as it is yet noticeable to us. They are so far removed that they can traverse immeasurably great spaces before we can see their apparent positions displaced by a few minutes, and for the determination of this so small a displacement the observations of ancient stargazers seem to be subject to too many other errors to be fit for a comparison with recent ones, a fact that should sufficiently be known. Meanwhile, the effort to undertake the investigation was nevertheless rewarded.[1]

What you, Sir, still consider arbitrary in my letter, concerns the partition of fixed stars into separate systems and you have given me for this some reasons which I tried to present to myself as clearly as possible. The first of these reasons is the analogy which when considered generally one can extend very far in the study of nature. For everything in nature [136] is arranged along general laws. You, accordingly, make of the whole world-edifice one whole system.[2] This you divide into single parts and each of these again into smaller parts until it comes to the solar systems, and from these to the systems of planets which have only a few satellites around them. Here I grant you that the earth with the moon, Jupiter and Saturn with their satellites, constitute the simplest systems, that the sun with all its celestial bodies is a notably greater system, that such larger systems are around each fixed star, that finally the whole world taken together is a whole, or the most complete system. The idea which we make of a system and the knowledge which we have of the world visible to us demand that everyone should grant this. Accordingly, the question truly is whether we are not making an astonishingly great jump when we advance forthwith from the solar system to the system of the whole world-edifice, and whether the fixed stars themselves should not be arranged into classes and these classes gradually into still more general classes. For otherwise we would have only three levels [of organization]: the system of each planet, the system of each sun, and the world-system. But if instead of these three levels there were innumerable levels, if these three levels did not make complete enough the subordination of systems, if there failed to exist a [satisfactory] proportion between the number of planets and comets around a sun and the number [137] of all the suns, then we should unquestionably admit still more levels. I have no second thoughts on this, because a chain of three links seems to me too short and in all places where we find levels in nature there are more [than three].

Here you, Sir, started to investigate whether one does not find traces of such multiple levels, and the investigation leads you to the point that you divide the Milky Way into innumerable parts. Each of these parts you make into a separate system of fixed stars or suns. All lie in one plane and this goes through the Milky Way. Our sun lies in one such system to which you assign all stars that do not properly belong to the Milky Way, or to those systems which that streak of the heavens presents to us. You separate these more distant systems from ours by a considerable intervening space, and the reason which you give for it is that the Milky Way is clearly distinct from the remaining parts of the sky. This is also the arrangement of the world-edifice visible to our eyes, and you say that in this representation there is still much which is arbitrary. I took pains to locate this arbitrary [element], because I was desirous to form a lively notion of the whole interconnection and to follow you, Sir, in such far-reaching conclusions. Please judge how far I succeeded, and [138] to what extent you would find prompting in it to go further.

To me the chief question seems to come to whether the fixed stars, which we see in the Milky Way through telescopes, are at least as distant from one another as the nearest stars are from our sun. If this is so, it becomes immediately clear that they ought to lie behind one another in indescribably long [many] rows. I take, for instance, two similar stars in the Milky Way which appear to be separated by one second. I posit that they are equally distant from us, and I have therefore an equilateral triangle whose two longer sides form an angle of one second, the shorter side being the distance of these two stars from one another. Trigonometry shows that each of the longer ones has to be 206,265 times greater than the shorter. The latter is, however, at least 500,000 times greater than the distance of the earth from the sun.[3] Therefore these stars should be 200,000 times 500,000, or 100,000,000,000,000, that is, a hundred thousand million times farther from us than the sun. Since I cannot imagine that we should nevertheless be able to see them, I rather conclude that the stars in the Milky Way are either closer to one another, or they should lie in long rows behind one another. The former case does not seem to me convincing. I simply cannot find any reason to put all stars in the Milky Way at the same distance [from us], and even if I wanted [139] to do so I would have to assume that they were very small and therefore we would not be able in all likelihood to see them because of their great distance. On the contrary, I see all the fixed stars as so many suns[4] whose light and heat should serve many millions of dark celestial bodies. This, however, demands a great sphere of influence in which such planets and comets might move and therefore also a great distance for the fixed stars from one another. You see, Sir, I am using your basic tenet about the habitability of the world in order to determine the distance of fixed stars and to place them apart from one another across the immeasurable space. I wanted to leave no fixed star for the mere purpose of being a spectacle, but wanted the light, the heat, and its gravity to serve the purpose of letting a whole system of celestial bodies wander around it and benefit thereby. Means without purpose seem to me good for nothing and I exclude them from the world. Each fixed star must serve similar purposes to which we see our beneficent sun dedicated. It has an area commensurate with its majesty and each fixed star deserves to be no poorer or simply unused. Its light and heat are means and these ought to be utilized just as we make use of the light and heat of our sun. To each fixed star I therefore grant a considerable area which stretches as far as its sphere of influence does. Consequently, I must certainly set the distance between two fixed stars at about as great [140] as that of Sirius from our sun[5] and the fixed stars must lie like rows behind one another.

Up to this point I believe I followed you, Sir, in an orderly fashion and I only arranged the matter according to my ideas. But you have gone much further and I must tell you that I found it somewhat difficult to make still further steps from there. I do not fully understand that I must extend these rows of fixed stars innumerable times farther through the Milky Way than outside it, for I am unable to imagine that outside, where the sky appears almost empty, instead of bright suns there should only be dark and invisible celestial bodies, or that such empty space be filled with sundry hyperbolic paths of comets. Should I posit outside the Milky Way many times shorter rows of fixed stars, then your conclusion certainly follows that the entire system of stars we can see must not be spherical but flat, and the Milky Way should be its ecliptic,[6] so to speak. I viewed it therefore as a very flat cylinder, or a spheroid, and when I took for its width a row of a hundred fixed stars, I had to assume for its length a row of millions,[7] so that the Milky Way, which is represented by this spheroid, might appear to be filled densely enough with stars.

[141] I did not find much difficulty here. On the other hand, I could not immediately perceive why you, Sir, leave an intervening space between the Milky Way and our [visible] fixed stars. You gave me no other reason for this than to say that the Milky Way is very distinct from the remaining parts of the sky, that is, the break comes sharply into view. I would have preferred in that connection that you had elaborated on the matter somewhat longer. The premise seemed to me too far removed from the conclusion, and in order to find the intermediate proposition, I had to solve a problem which appeared to me very puzzling. I am far from being an Oedipus.[8] Meanwhile, I mustered my strength to make the matter comprehensible to myself. I know how fast you proceed, but I should never be lacking in proofs to show you that I take pains not to fall behind.

I started accordingly to fill this space, which you, Sir, left empty, with stars as densely as the other spaces, and in order not to exert too much my power of imagination I drew the whole matter into a small picture and tried to find for myself an illuminated place over which the lamps were placed in much the same way in which the fixed stars are in the plane of the Milky Way. I started with the ground and imagined on it a few thousand rows of lamps that stood at the same distance [from one another]. In equal height over this I put an equally extended layer of [lamps] and above this [142] still another 100 layers. Thus the whole air in the entire place became filled with lamps and the place was more dazzlingly illuminated than it could ever happen. Then I put myself inside that space and watched what it looked like. When I looked straight up or down I found only 50 lamps. But the more I turned my eyes toward the horizon, the more lamps I saw in a row. Their number, however, increased gradually, first slowly, afterwards, however, more strongly, just about as the secants of an angle increase. But I nowhere saw a break as we see it in the Milky Way. I followed therefore your conclusion, Sir, and I let the nearest lamps stand around me, but between these and the more distant ones I eliminated a big circle [of lamps] and looked once more at the illumination. Now the lamps which remained very near to me were recognizable in magnitude and distinctness. Those farther away were more simply, unclearly, and densely together and the break came sharply in view.[9]

Thus I made a vivid picture to myself of your conclusion which certainly exercised my power of imagination, but it was not yet as much strained as when I wanted to present to myself the true illumination which obtains in the firmament. But now I can do even this because in such a way I began to perceive your conclusion more extensively. Certainly, the density of stars in the Milky Way would not appear to diminish abruptly, but would [143] appear to decrease gradually if there were no wide intervening space between it and our system of fixed stars.

From all this I also conclude that you, Sir, have stumbled on the most natural of all explanations given previously of the Milky Way, because you give many reasons which are closer to probabilities than to mere possibilities. In ancient times one could at most guess whether this streak consisted of the fusion of the light of smaller stars. The telescopes have put this beyond doubt, and one then has only to try to see what can be made of this. In all speculations that one dared to make about the Milky Way the question still remained unexplored why only this streak appeared so full of stars.[10] This is just the question which I proposed to you, Sir, in my last letter. If one posits that the stars there are very closely together, then the question arises: why only in this streak? If one posits that they have stronger light and that those outside the Milky Way cannot be seen, because of their weaker light or simply because of its lack, then the same question still arises. Such explanations are not only arbitrary possibilities, but when one wants to give their reasons one has to say that it is so because it is just so, that is, one cannot yet find any reason.

So far I readily agree with you Sir, that [144] your explanation seems to come very near the truth and I hope that you will find still more reasons for it. How much would it please me if the explanations which you gave about it would be some service to you! In particular, I beg you to see to what extent your initial assumptions might be brought to as much clarity as may be required. In my opinion the world-edifice that comes into our view is made flat through a necessary conclusion as soon as it is sufficiently demonstrated that the fixed stars in the Milky Way lie behind one another and that one is as far from the other as the nearest fixed stars are from us. I tried to deduce this from the utility which the light of fixed stars must have, just as much as the light of our sun has it. Our sun illumines also through the nights of the inhabitants that are around Sirius, Arcturus, and other fixed stars. But it has a still closer utility which is spread upon all celestial bodies of its system and therefore also upon our earth. If therefore I place around each fixed star a system of bodies which must use its light and heat, then I must assign to the fixed stars a similar distance, and this distance increases with the size of the star because its sphere of influence is proportional to its mass. But that the stars in the Milky Way hardly fall behind our sun in size and brightness, you have, Sir, inferred because in spite of their great distance they are still visible. It will [145] please me very much if you were to try to make this proof even more stringent. Your considerations on this far-reaching and still quite undeveloped matter have all likelihood, and you give me, in reply to the main question which I submitted to you, a reason which has a close connection with several others. It becomes very clear to me that in the sky there ought to be enough empty places so as to leave room for movement, because movement is essential in the world. You show this also from the analogy in our solar system where no several celestial bodies are in one and the same orbit. You show that thereby the harmony of the world becomes more complete, because it then extends in time and space. And now I no longer miss the symmetry in the apparent position of the fixed stars, because I clearly grasp that our solar system would not appear more orderly were all its bodies visible. The degrees, through which the systems become always greater, more prominent, and more majestic with respect to time and space and grow into the whole world-system, are completely in accordance with the analogy which we see in the whole of nature. I also divide with you, Sir, the Milky Way into innumerable smaller systems of which perhaps some that are separated from others might be distinguished. Since all these systems are in motion and must have their definite orbits, I will no longer be taken aback by the irregular shape of the Milky Way. The order consists in the arrangement of these orbits [146] even though it may remain hidden to us for a long time because we lack number and measure for the determination of so enormous spaces and times. Meanwhile, one can safely conclude that this arrangement is not in the least less real, for order and perfection necessarily extend throughout the whole world.

Although I admit that all these systems lie in one plane and I also state that this plane has a considerable width, one question still remains unsolved for me, because it occurs to me that the position of the systems of the Milky Way must have another arrangement than the celestial bodies of our solar system. You will recall, Sir, that you prove with good reasons that there could be more celestial bodies around our sun, if their orbits do not lie in the same plane but have all possible inclinations with respect to one another. Also, I would almost be ready to conclude that the number of systems that constitute the Milky Way is by far not the largest that would be possible. I know well that this conclusion does not yet necessarily follow, because it is not known where all these systems shall turn to in their orbiting as time goes on. It would be inconceivable to assume that we live just at the time when all these systems seem to lie in about the same plane. Or did you wish to compare the whole Milky Way in respect to the whole world-edifice only with the system of Saturn and Jupiter whose satellites are also in the same plane? You readily [147] speak to me of still other Milky Ways which are outside the one visible to us and which taken together with ours must constitute a greater system. I suspect that you would look at the faint light one sees in Orion as such a Milky Way which would be the one closest to ours. But I see on this point so little likelihood of bringing out from general considerations something more definite, that it will already be satisfactory for me to have but a general overview. Still I think that one can make here more than the first step, and I see what you, Sir, write to me about the position of our system as proof of it. Even though you offer your thoughts on this point only as a hypothesis, the effort to derive such consequences in greater amounts will always have its rewards. A hypothesis will in the end become a truth, when all phenomena let themselves be derived from it in a natural and in an obvious manner, when all these consequences are connected with one another and with the general reasons, in short, when that hypothesis is consistent in all its parts with itself. When in the edifice of truths there remains somewhere a gap and one finds a system of thought which fits exactly in that gap, then it becomes very likely even though one cannot quickly connect it with all the remaining truths. One comes closest to the truth when new phenomena can be inferred [148] and foreseen from it. You know, Sir, that it so happened with the Copernican world-edifice which no astronomer doubts any more. Since you try to make the whole world-edifice completely Copernican, you build on similar grounds and I see it as a good omen that you are having such success in the process. It will satisfy me if I can only provide one handyman to your edifice. You know how little will be lacking in me of that well proven readiness to co-operate on it, and how much I always make it my concern to show you with ever new proofs that I persevere as long as I live, Sir, etc.

[149]

TWELFTH LETTER

What I have expected, Sir, of your friendship and begged for in my last letter, you afford me to my greatest satisfaction. I readily grant you that I have made conclusions about the position of fixed stars which were rather a result of insights than of deliberate reflection. You have therefore obliged me very much by taking the trouble to bring these conclusions into an order and to test their good and weak points, and now I clearly see how far I get with it and where I fall behind. I still must tell you how I came to this chaos of thoughts as far as I can remember.

In a clear night[1] I sat at the window and as the objects on earth put aside for the next day all their charm to draw attention, there still remained for me the starry sky as the most worthy of contemplation among all showplaces. You know how many hours I sacrificed to it since childhood[2] and how little did the habit succeed in weakening the pleasure in this contemplation, or in making it a trite daily affair. It may be that the realm of stars unfolds always new rarities, or that the variety [150] in it is inexhaustible, or that the glimmering light of stars has in it something very pleasant and charming for the eyes, or finally that an astronomical eye never gets tired because it finds a constant *plus ultra* and because the sky gives him ever new material for enraptured astonishment and for reflections which the quiet of the night helps to gather and makes more lively. With me all these reasons unite when I consider these shining lamps in the temple of divinity. Then I take the wings of light and soar through all spaces of the heavens. I never come far enough and the desire always grows to go still farther. In such reflections did I present to myself the Milky Way. I once more felt astonished over the army of small stars in that arch and missed them outside it. The stars, I thought, are not so close together as to touch one another. They ought to lie behind one another, and the rows of stars should throughout the Milky Way be many times longer than outside it. Were they equally long in all places, the whole sky would have to shine as brightly as the Milky Way. But outside that streak I see but almost empty spaces. In brief, the edifice of fixed stars is not spherical but flat and very much so.

Here I remained during the first night with my astonishment and also with my last conclusion. [151] Only somewhat later did it strike my eyes that the Milky Way is very distinct from the remaining part of the sky, and then I started to push it farther out and to admit the empty space which you, Sir, brought so neatly into a sensible picture. I put all the remaining stars into one system, but the Milky Way remained undivided for many days.[3] Later I was puzzled that I did not stumble on this right at the outset, although the division clearly shows itself. But apart from the fact that unnoticed things almost constantly emerge before our eyes, there still could be another reason. The section of the Milky Way which is split up did not stand opposite the window. But since I still was desirous to see whether I could further develop these thoughts, I surveyed the Milky Way once more, especially its shape. And it was then that I first concluded that it had to be divided into single parts and that each part was similar to the one in which we find ourselves. So arose, almost unnoticeably, my systems of fixed stars.

It may be that gradually more thoughts occurred to me about the Milky Way because I have not yet given up the subject. But in all likelihood it must have happened very slowly because we have no power over our insights. These will be prompted most when other occasions are forthcoming, and you, Sir, have provided me with the [152] most desirable of such occasions when you have posed me precisely this question, and now in your last letter you have brought into order my still unclear conceptions and properly arranged them. I accept with pleasure your proposition and I will now strive to explore what is required for a complete proof, what is still lacking in it in my case to make it complete, and how far the gaps in it may be filled by [considerations of] probability.

First, it is unquestionable that my hypothesis will necessarily be either approved or rejected when we shall have a means to measure accurately the distance of each fixed star, because then one would know for certain their position which I tried to determine by other considerations. The yearly parallax of the earth is for this purpose very insufficient,[4] and therefore the means is cancelled which geometry offers us in other cases. The law of gravity and the motion of the fixed stars gives us another means, but this motion is so small and its features so undetermined that we shall achieve nothing with it even in many centuries.[5] For the time being this can be applied only in the case of the bodies of our solar system. The only means that remain is the light and the magnitude of the fixed stars, two items which must be considered both in themselves and in their visual appearance if one is to derive the distance of the stars. [153] A fixed star appears to us greater than another, not only the greater and closer it is in fact, but also the brighter it is. We see this at night also with lamps whose apparent size decreases[6] with distance and which also appear bigger when they are brighter. The reason for this is due to the opening of the pupil, to the unclarity of the image on the retina, and to the scattering of light upon it. The fixed stars appear as bright points[7] only through the best telescopes because they eliminate the scattered and spurious light. For similar reasons also one star appears brighter than the other, but particularly when it is indeed brighter. It may furthermore be conjectured that the light of the fixed stars weakens somewhat while it reaches us. That it weakens in our atmosphere is unquestionable, we see this also about the sun. The light can, however, be held up in part in the ether,[8] pure as this can be. The atmosphere of the sun necessarily holds up much of its light.[9] There may be fixed stars which have even greater atmosphere, and the space between them is far from being completely empty. For this reason the farther a star is, the more dimly it must shine. Were, however, the distance the same with each, this reason would disappear because the light from all of them would be weakened in a [similarly] proportional manner.

If all fixed stars, considered in themselves, [154] had the same size and the same brightness, it would necessarily follow that the [apparently] smaller ones had to be more distant. Were they, however, equally distant, then one would have to assume that the smaller also had a weaker light. For the number of stars of the first, second, and third magnitude increases approximately like the square of magnitude. Of the first magnitude one counts 18, of the second 68, of the third 209, of the fourth 453, etc.[10] You already see, Sir, that this progression fits the various distances much better. I would not state though that all stars are equally great and equally bright. For then I would not get more than 12 stars for the first magnitude, or 48 for the second, or 108 for the third.[11] Some of the nearest stars can be smaller and darker, whereas many of the more distant ones can be brighter and bigger, and in general such a variety better suits my hypothesis because I put the stars in motion, and so they can always change gradually with respect to position and distance.

This motion, Sir, you have admitted as necessary in the world, and I do not see what objection one could make against it. The perfection of the world and the law of gravity, which keeps the whole world together, show that it is present in all places. I concluded from this without any hesitation that the fixed stars ought to be unevenly [155] distant. An equal distance would be too uniform. The central forces do not permit it. Because the fixed stars must orbit around a center, they cannot move on the surface of a sphere. They would cross one another and hit one another. The world-systems must be maintained, and therefore no orbit can cut across the other. I cannot very well do otherwise but leave around each fixed star a space which is proportional to its sphere of influence. One can also show from geometry[12] that, for instance, one can adjoin to the sphere of influence of our sun twelve other suns that have the same size. With that I would have but 12 stars. But they are innumerable and therefore the rest of them must lie along rows farther out. Each star has a sphere of influence because it is heavy. Each must furthermore have space for its orbit. Everything leads me therefore to the conclusion that they spread out through the whole cosmic space. If I still add the consideration that in the sphere of influence of each star there is a system of millions of comets and planets that use its light and heat, so that it could be useful in a more immediate way, then I do not see what one could still demand in the way of proof unless the autopsy itself.

Meanwhile, I will yet make this consideration which is necessarily conclusive, because one can as little deny gravity to the fixed stars as to our sun. If you take two stars whose apparent [156] distance amounts to only one second [and] if they are equally distant from us, then their true distance from one another is the 200,000th part of their distance from our sun.[13] The nearest star might be 500,000 times farther from us than the sun.[14] When I therefore give to both these stars such a distance, they will be $2\frac{1}{2}$ times farther from one another, [and therefore] they should have either long since fallen upon one another[15] or, if such is not the case, they should have an orbital motion around their common center [of gravity]. This orbiting should necessarily be visible to us through telescopes because they should not have too long a period at all. One should therefore observe a steady change in the position of these stars, one of them should be now in front of, now behind the other. About all this one does not observe anything at all, and the change in the position of the fixed stars since Hipparchus' time down to ours can hardly be ascertained.[16] From this I necessarily conclude that such stars must have a very different distance from our sun. [17] You can easily guess that in the Milky Way everything would swarm if the stars in it did not lie behind one another in inconceivably long rows. One would also see the same in the so-called nebulous stars.

Add to this, Sir, that we have no reason to be so thrifty [157] in respect to space in the world. For should space be dealt with thriftily, the gravity, the central forces and in general all motion would be eliminated. Our earth should then stay still in one place, and its orbit would be filled with other planets. But in this way the most beautiful and the most varied [factor] would disappear from the world. For its perfection motion is much too necessary; where motion is, however, there room must remain. And it is clear in itself that the central forces, the orbits, the solar systems around each fixed star demand motion and space.

Here you have, Sir, the proofs which I was able to produce on behalf of the first proposition, namely, that the fixed stars and especially those that are in the Milky Way lie behind one another, and that their distance from one another is very considerable and ought to be proportional to the sphere of influence. The conclusion which I draw from this is that outside the Milky Way there are no such fixed stars at such a distance, and precisely because the remaining parts of the sky seem to have empty places. Since I consider these empty spaces necessary for the orbiting of the systems of fixed stars, I do not hesitate to take them for being really empty, and into such a night I would be putting with no justification dark and unilluminated systems of celestial bodies. [18] There must remain room for motion, slow as this may be.

[158] If I now conclude from this that our visible star-edifice is not spherical but very thin and flattened, this conclusion is then nothing more than an application of well-known geometrical propositions. You have, Sir, made the matter so clear through your illuminated place that I have now no need to spend much time on it. You have thereby equally explained the separation of the Milky Way from our system, and I do not believe that you find difficulties when I distribute the whole Milky Way in such smaller systems which lie in it behind one another. I do not determine the number of such systems, but in all evidence it must be indescribably great. I have concluded this already in my last letter from [the principle of] analogy, as I started with the system of Jupiter and proceeded from this to the system of our sun, and from this to our system of fixed stars. Number, space, size, and time keep increasing with one another.

Then as I separate our system of fixed stars from those that appear in the Milky Way, I leave between these a similar intervening space. They lie in about one plane. Therefore I cannot assume more than six that are immediately around us if I set their distances from one another [to be] about the same. All the rest I must remove farther and farther out. The apparent diameter of the next systems of the Milky Way can [159] hardly be equal to its mean width and therefore not more than 10 degrees.[19] If I take the true diameter of such a system as a yardstick, then it has to be laid off six or more times to find the distance of such a system from the next bordering on it. I think, however, that the apparent diameter amounts to less than 10 degrees and perhaps hardly to 5 or 6 degrees, and I conclude from this that a system will be removed from the nearest 10 to 12 times more than it is wide.[20] Since, however, the Milky Way is quite compact, uncounted such systems must therefore lie behind one another to cover each intervening space which the next ones would leave empty.

Yet their number is definite and the Milky Way must have its limits somewhere. I have derived this in my last letter from [the principle of] analogy, since I have posited still uncounted other Milky Ways outside that streak, because it does not seem likely to me that the world-edifice is so confined, though on the contrary just as little do I grant that it is infinite.[21] A great deal could be said if you, Sir, ask, whether the pale light in Orion, which Derham considered to be an opening in the *coelum empyreum,*[22] is not the nearest of such Milky Ways, and I will not now address myself to the point that a change has been observed in its visible shape.[23] It appears to be too far removed for one to be able to see it always equally clearly [160] through our air and through any telescope.

One can perhaps conclude most correctly from the apparent form of the Milky Way that it has narrow confines. Were it immeasurably extended, it should appear to us as a great circle of the sphere. But this is not so. Its apparent form is rather an oval. Its center stands from the North Pole 35, from the South Pole only 25 degrees away. On the other hand, it cuts the equator into two fairly equal parts. So does a ring appear to the eyes when one is located both outside it and away from its axis. Our system of fixed stars appears to lie not only somewhat outside the plane of the Milky Way but also closer to its periphery than to its center.[24] The smaller irregularities in its apparent figure do not allow any exact determination.

Since then there occurred to me another means by which one could at least try to determine whether the fixed stars of our system, and therefore also our sun, gravitate toward a body which lies in the center of that system, or whether they merely move around the common center of gravity of the whole system. Would the first be the case and were our sun gravitating considerably toward such a body, then the planets also should have such gravity. It is in such a way that the moon, [161] for instance, is attracted not only toward the earth but also toward the sun, and the astronomers complain about the smaller though still very noticeable irregularities that arise from this. To such irregularities even the planets ought to be subject, and therefore also our earth in its annual orbiting. Therefore the question is whether one would notice such yearly anomalies through careful observations, whether the observed position of the sun would coincide with the calculated one within seconds throughout the whole year, whether the retrogradations of the aphelia and nodal lines in the orbits of planets could be derived from it,[25] whether the change in the obliquity of the ecliptic could not be in part traced to it. If the plane of the orbit of the remaining planets is displaced, one has no reason to leave the plane of the earth's orbit steady, [just] because it is nothing less than the yardstick of the rest. The inclination of other orbits is reduced to that of our earth, or the ecliptic, only because this is most convenient for our calculations and because one always has to compare them with one another.

If one could here determine something, it would unquestionably follow that the orbits of planets are perfect ellipses as little as is the moon's orbit,[26] that the sun's distance varies in a double manner as does that of the moon. One can, of course, easily recall that [162] the astronomical tables often give the position of the sun in such a way that the observation differs from it by almost a whole minute.[27] It needs therefore only to be seen whether this happens at regular times and results from this cause.

From this you will see, Sir, that there is also a means to prove by observations my considerations on the orbiting of our sun, provided its gravity toward the common center of its system is considerable. One could in such a way find the area of that center, because all planets, and especially our earth, gravitate toward it, and the smaller anomalies must point to it just as those of the moon point toward the sun. In my last letter I raised the question whether there ought to be in that center a dark body of enormous mass, or whether one ought to take in that connection only the common center of gravity of the system of fixed stars. Perhaps this question can be clarified through such observations, or perhaps one will at least be able to conclude from them how much gravity each planet, and therefore also the sun, has with respect to that center. This gravity is certainly very minute, because the irregularities in the motion of the earth, which might therefore be affected, are not yet observed in a satisfying way, but still ought to be first determined and investigated through very exact observations. With the moon one found much earlier similar deviations which are caused by the sun.[28] For the computation [163] never wanted to coincide exactly with the observation, and the deviations struck, so to speak, the eye because they amounted to several minutes. With the earth they seem to be smaller than one minute, as I have already remarked. Exactly the same can be concluded from the slow motion of the aphelia and nodal lines,[29] especially when one computes their advance not from the equinoctial points but from a fixed star.

I could pursue so far, Sir, your suggestions to elaborate the doctrine of my systems of fixed stars. I have not at all reached your clarity, and I believe that I have still left much in confusion from the first chaos of my insights. I will therefore be intent on accumulating still more conclusions until I see what you find worth noticing about the material available so far. Do not think that I am dashing ahead of you. You always show me how far the road really goes and to what extent I should distinguish mere guesses from firm conclusions. I readily retract all that you do not find consonant with your thoughts because I would rather sacrifice all my systems for this precious harmony. This should forever remain the reason for my satisfaction and for the perfect friendship and loyalty in which I persevere, Sir, etc.

[164]

THIRTEENTH LETTER

How pleasant would it be for me if I could provide for you, Sir, the occasion to bring to all completeness your considerations on the ground plan of the world! I hope you will see from all my letters that this matter arouses my attention more and more and leads me to serious reflections. I feel every satisfaction about the view of the starry sky which you so vividly describe and I wish I had started earlier looking at this splendid stage with astronomical eyes. But I will all the more eagerly make up for this neglect and try to put myself at least in the position to submit to you suitable questions. These questions worry me no longer, because I realize that they are not so terrifying as those which one raised about the effect of comets more out of idleness than for serious reasons.

You have obliged me very much, Sir, by describing how you arrived at your system of thought. Although you are much intent on seeing your first insights about it as a chaos, I still must tell you that I find nothing in them that would be different from your other fortunate insights. I know that [165] you always fall back on the most probable [case], and when it comes to the proof there does not remain much for improvement. Your system of thought is so arranged that I have unwittingly matched only those ideas of yours which indeed fit together. You feel each gap that still remains there, and each new occasion serves you to fill one or several of them because you know how to assign to each new experimental evidence its place in the system.

As to the various distances of the fixed stars I have now no considerable doubt, or rather no doubt at all, when I take together the reasons which you, Sir, have given on this point. It has already been long considered as something simplistic [to think] that the ancients had affixed all the stars to the surface of a sphere because they wanted to explain their *primum mobile,* that is, the daily rotation of the firmament.[1] In fact, it had to seem fairly simplistic, if this rotation had not been apparent and if it was not to be ascribed to the earth as much as was the slow change [motion] of stars around the pole of the ecliptic. It is now sufficiently proven that the entire motion, which the whole firmament seems to have, derives from the earth alone, and in general also those motions that follow a uniform law and recur each year or each month, such as the aberration of light and the nutation of the earth. Since this [uniform motion] was the only reason which the followers of Ptolemy gave for [166] nailing the fixed stars on a sphere and for setting their distances equal,[2] this reason now completely vanishes and no further reason remains.

On the contrary, we have innumerable reasons for setting these distances to be unequal and for putting the stars in long rows behind one another. First, there are the suns with legions of celestial bodies around them to which they give light, and heat, and changing seasons and make a dwelling place out of them for living creatures. Such a system demands space as great as is the sphere of influence of each sun. This sphere of influence has to be necessarily very big and not at all smaller than that of our sun, because the stars are seen to shine in spite of their great distance with such a brilliance. With this goes a measure of light, density, and size, much as is the case with our sun, and therefore an equal sphere of influence. No sun can be in the sphere of influence of another without moving around it, or both moving around their common center of gravity, and this motion must necessarily become visible even to us in a few years.[3] From this you conclude, Sir, and rightly so, that in the Milky Way and in the nebulous stars everything must be swarming.[4] This is a matter which in a short time will be cleared up through observations and you can safely refer to these as you can to various other points. About the bigger stars it is evident that [167] in many years their relative positions did not change noticeably and one has to see from Ptolemy's catalogue of stars, compared with the more recent ones, whether such a change can be demonstrated.[5] This method of letting the decision in the matter be based on visual evidence you have already presented in your previous letter.

Cassini considered it to be a consequence of the annual parallax of the earth that the first star in Aries appears double in certain times.[6] To ascertain this, one should investigate whether this phenomenon occurs again each year, or whether the varying transparency of the air does not permit that one should see this star always as double or as really two different stars, and whether they have their position changed with respect to one another. The same remains also to be investigated about the middle star of Orion concerning which Huygens had found that it consists of 12 smaller stars.[7] In such a way it might be clarified whether in fact there are fixed stars which are in a common sphere of influence and orbit in a short time around the center of their gravity.

But until this is cleared up you may, Sir, safely stay with your system[8] because I do not expect anything of that sort. On the contrary, the other observations, which you have made with respect to the planets and [168] especially also about our earth, seem to me of greater usefulness, because from it various points can be determined about the position of the sun in our system of fixed stars and about its motion, and what you say about the smaller deviations in the Tables of the motion of the sun[9] certainly seems to give the hope that this investigation will not be fruitless. Even more excellent would it be if the displacement of the aphelia, of the nodal lines, and of the inclination of each planetary orbit could be derived from it.[10] The comets, too, can certainly contribute something to such very small displacements when they come close to a planet; but this would not be general, and especially it does not seem that such great planets, like Jupiter and Saturn, might be affected by this at all. Therefore the ordinary feature of these motions remains all the time dependent on the center of our system of fixed stars and has to be derived from it.

In this respect assume, Sir, each planet to be subject in its orbit to about the same anomalies which the moon suffers because of the sun. You consider the sun as a body which gravitates toward the center of the system of fixed stars and you look at the planets and comets as its satellites. As much as the sun gravitates toward the center, so does each planet. Therefore the planet must certainly move now faster, now slower than this [169] would happen if it gravitated only toward the sun. But when I push this comparison further I find a difficulty which I cannot easily remove. It concerns the direction in the motion of the nodal lines. This moves in the moon's case against the sequence of the signs of the zodiac, but in the case of the planets it follows this sequence. What do you think, Sir, whence might this come, or how might it be explained from the universal gravitation of our solar system toward the center of the system of fixed stars?

Then you have, Sir, in your previous letter concluded from seeing all around us scattered fixed stars that our sun should not be near the borders but rather near the center of the system of fixed stars to which it belongs. This unquestionably makes its motion faster and its gravity toward the center greater, and all the more this should also be noticeable with the planets. But since the anomalies arising from it are very minute, you must assign to our sun a very great distance from that center, especially if a dark body of enormous mass should be there, toward which even the outermost suns of the system would have some gravity. Then, in fact, such a body would not be of much use if its diameter were not almost as big as the sphere of influence of our sun, or the whole [170] extent of our solar system. If the fixed stars outside the Milky Way could be partitioned into more than one system, then I would readily put such dark bodies there, where the astronomers saw new fixed stars and where others have vanished. Then the appearance and disappearance of fixed stars could be explained in much the same way as the eclipses. But it will always be better if one first investigates with careful observations whether the gravity of the sun toward the center is considerable, because it would then be possible to determine somewhat more closely which of these possibilities is true.

Since I consider according to the proofs, which you, Sir, have given, as morally certain the motion of the fixed stars and their dissimilar distances from us, I have tried to calculate their velocities. I first assumed that the nearest fixed star has moved $\frac{1}{4}$ degree since the time of Hipparchus or, using a round figure, in 2000 years. This I could easily assume because Ptolemy's catalogue of stars will not easily be made more exact.[11] I then assumed the distance of the earth from the sun as a yardstick and I posited that this fixed star is 500,000 times more distant. By considering this number as the radius of a circle I found that about 4000 such yardsticks[12] would cover $\frac{1}{4}$ degree which the star had to traverse [171] in 2000 years. Therefore it would traverse in one year two such yardsticks, or a space [distance] which is as great as the diameter of the earth's orbit. Since this diameter is only $\frac{1}{3}$ of the earth's orbit, I find therefore that the velocity of such a star is three times smaller than that of the earth. This calculation is correct if the sun remained at rest during that time and the star moved around it in a circle. When, however, I assume that the sun also moves, then only a relative velocity is the result, which is bigger or smaller according as I change the direction of the path. For a similar reason the planets appear to us to be now moving forward, now backward, now standing still. There can be therefore fixed stars of the first magnitude that have not changed their position noticeably and others for which the change of position is more noticeable.[13] Since, however, I put the most noticeable change at $\frac{1}{4}$ degree, I must take also the circumstances which make it greatest, and these obtain when the sun and the stars have an opposite direction. This, however, brings the velocity of the star to a half and accordingly I would put it at six times smaller than that of the earth. Because of this the gravity of the star toward the center of the system becomes very small, if I consider how great are the circles and ellipses in which the stars move around this center. Meanwhile, this gravity is many times greater than that which the two nearest [172] fixed stars have toward one another or toward our sun. For since the velocities of planets are approximately as the square roots of their distances from the sun, a star which is 500,000 times farther away than the earth has to move 700 times slower. According to the foregoing considerations the star in question moves only 6 times slower and therefore 120 times faster than if its central forces depended on the sun. A greater velocity invariably requires a stronger gravity if the body is to retain itself in its orbit. I would therefore certainly conclude from this that the gravity in the center of the system of fixed stars must be very strong. It is, however, superfluous to go on with this incidental calculation, because the proposition on which it is based, namely, the magnitude of the displacement of fixed stars since Ptolemy's time, must first be established.[14]

Still another question I must submit to you, Sir. Since you separate the systems in the Milky Way from ours by a 6 to 10 times larger intervening space and thereby remove them so much, I cannot understand how we can distinguish with telescopes the stars that are in that streak. You must place the nearest of these at least several hundred times farther than Sirius. That we see in that streak only a diffused light with the naked eye, I can still very well derive [173] from the fact that the images of these stars fuse on the retina and therefore must necessarily appear unclear to us. Our atmosphere, too, can contribute something to this, and perhaps the same is done by the scattering of light in the ether. How exactly then should the telescope remove all foreign light if the stars are to appear to us at such distances as bright, clearly distinct points?

Moreover, I see very well that since the number of fixed stars increases approximately as the square of their magnitude, this law can more readily be explained through their varying distance because the surfaces of spheres increase in such a proportion. In addition one could conclude from this that the fixed stars in our system are still fairly uniformly distributed because their number increases incidentally as the surfaces of spheres whose radii are 1, 2, 3, 4, etc.[15] The motion of stars allows no greater uniformity because they change their places gradually.[16] Therefore I do not believe that this law will be extended much farther because the telescopic stars are in some places of the sky much denser than in others. Especially is the area around Orion occupied to an astonishing degree with great and small stars. The constellation of Canis Maior and the Argonaut also seem to be equally rich in stars. They lie in the same plane with the Milky Way, [174] and if our sun is noticeably far from the center of the system, then I must strongly suspect that these constellations lie beyond the center because there the row of fixed stars is longer [deeper], so that we can see more of them lying next to and behind one another.

In such a way would I take care of the position of our sun in the system of fixed stars to which it belongs, as you, Sir, work on establishing the position of this system in relation to all systems in the Milky Way. I have observed the shape of this streak on the vault of heaven. Although it is not completely circular, it still deviates unevenly from the poles and thus, as you remark, Sir, it represents an oval. I strongly believe that we are closer to that part of the Milky Way which cuts across the colures of the Ram, where it also has a double width and appears to be split. The smaller axis of the oval seems to go through this part. It stands just about opposite to the constellation Orion from which I wanted to move our sun farther away so that the row of stars lying behind one another may be all the longer [deeper].

To make all these determinations there are still lacking for us number and measure which presumably one will find only gradually as time goes by. It is, however, enough to know the position of the system of stars [175] in a cursory manner and to determine accordingly the place where we find ourselves in the world-edifice, even though this should be done in a negative manner. For instance, our earth is not in the center of the solar system but revolves around the sun. The sun is not in the center of its system of fixed stars, but this center lies in the area of Orion and Canis Maior. Finally, this system is neither in the plane nor in the center of the Milky Way, but somewhat protruding above it and closer to that part of the Milky Way which goes through the colure of Aries where it has a double width.[17] Where the Milky Way lies, if we are to compare it with innumerable other Milky Ways, cannot be easily decided because they are invisible to us due to their enormous distance. So much is, however, clear to me that if the weak light in Orion is such a distant Milky Way, many more similar ones ought to be discovered through telescopes if one is willing to search for them.[18] That one has discovered the one in Orion is probably so because this constellation unquestionably is the most beautiful, and it enticed the astronomers to all the more diligent observations because of the crowd of stars in it. Although it shines most brightly on the coldest nights, I view it each winter with new delight and I think with pleasure of the evening hours which I have spent with you, Sir, in the contemplation of this splendid [176] area of the heavens. It seems to encourage both of us alike to look for something very important[19] there, and after the previous considerations I would let the center of our system of fixed stars as well as the center of the whole Milky Way lie in that area. How pleased would I be if this remark deserved your approval! Write to me, Sir, what you think on this score. I expect your reply with the greatest eagerness, because I hope to find in it the solution of the question which I submitted to you and which I will receive, as I do all manifestations of kindness that come from you with the recognition with which I remain obliged to you, Sir, etc.

[177]

FOURTEENTH LETTER

Your most treasured letter, Sir, is to me a joyful proof that you make great steps in advancing on a road through the firmament and in making further progress on so worthy a course. You feel the whole vigor of the words of Ovid:

> Os homini sublime dedit, caelumque tueri
> Iussit et erectos ad sidera tollere vultus.
> [He gave to man an uplifted face and bade him
> Stand erect and turn his eyes to heaven.][1]

In fact, there is nothing more proper for man than to look also where nature directs his eyes and become familiar with the place which the All-wise assigned to him in the world. Small as we find our body while contemplating this immeasurable edifice, our spirit becomes great when we get in the habit of extending it through the whole firmament and infer from such great things the greatness and majesty of the Creator. You know, Sir, in its full extent what this means and you grant to astronomy among all sciences the first dignity, because it is, and whatever serves it, the only thing that endures through all eternity, and the heavens shall declare to us the glory of God even after death and announce to us the traces of His handiwork.[2]

I turn with pleasure to the investigation of the questions [18] which you have proposed, and I will start with the one which you raise about the possibility of distinguishing stars in the Milky Way when we view it through the telescope. I readily grant you that this investigation is very difficult and that it rests on reasons which one has not yet sufficiently developed in optics.[3] Meanwhile, I will make another attempt to see to what extent I shall succeed. Here one deals with such propositions to which I will not attach proofs in full, lest I hold you up unnecessarily.

As long as we clearly see an object with the naked eye, it appears to us more or less equally bright regardless of its position and distance. I say *more or less*. For the variation in the opening of the pupil can change something there. But because one can always take this into account, I will take this opening as steady, and therefore I conclude that we see each object in its real brightness as long as its image in the eye is clear. In such a way the sun would appear smaller, but still equally bright if we looked at it from Saturn or from an even greater distance.

When, however, an object is so far removed that we can see nothing clearly on its surface, then two cases [can] occur. First, [consider] when [179] the apparent image is considerably large. In this case the light of each part fuses together, and we have on hand the average of all particular brightnesses. This average is again always the same even though the distance changes considerably. Thus we see a distant wall equally bright when it is illuminated by the sun at a given angle, while we may be closer or farther from it. And in the same way the moon would appear to us bigger but not brighter, were it closer to the earth. For although more light would reach the eye in this case, it would spread out more on the retina in a similar proportion. This is also the reason why we see objects as equally bright through convex and concave glasses.

Should, however, the object appear to be unclear and only as a point, then the situation is different because its image on the retina takes up a bigger space than it should if we saw it clearly [in some detail]. The rays scatter and if two such points are near together, the scattered light mixes and we see both points as one. Thus we see the fixed stars shine weaker than if we could see them clearly with the naked eye. For in this case they should shine as bright as the sun, and again regardless of their distance.

[180] We try to obtain this clarity through telescopes for which the same holds with respect to clarity as what I have said before about the eye, with the difference though that the telescope does not depend so much on the opening of the retina as on the opening which one has to give to the object lens. Therefore the telescopes do not present to us objects as brightly as these could be seen with the naked eye, if one could see them equally clearly, but they decrease each brightness in a proportional manner, and again regardless of the unequal distance. If therefore one wants to compare the brightness of planets with one another, this has to be done with telescopes which present them to us clearly.

From this I conclude that if we could see clearly through a telescope a fixed star as a round ball which, considered in itself, was as bright as the sun, its lustre would appear to us through the telescope just as bright as if we looked with that telescope at the sun. But since the true apparent diameter of the nearest fixed star is hardly $\frac{1}{4}$ tertia[4] of a degree, all hope is lost for seeing a fixed star perfectly round through telescopes. It always takes up a greater area on the retina than it should if we could see it clearly.

[181] If therefore I assume that a telescope magnifies 120 times, this $\frac{1}{4}$ tertia becomes 30 tertia, or $\frac{1}{2}$ second. I do not believe that the eye can see distinctly a point which appears under so small an angle. Also, the light that falls from the fixed star on so small a point of the retina is still always as strong as the one which would fall on it if we viewed the sun through such a telescope. Since it also causes a strong agitation on the nervous system of the eye, this agitation spreads to the neighboring nerves, and as a result the image of the star becomes greater.

Since telescopes serve to make each point clearer and to bring more together its image on the retina, it is clear why the fixed stars appear to be smaller, but also the more sparkling, the better is the telescope. According to the previous consideration the image had to be $\frac{1}{2}$ second. I assume that it expands because of the spreading agitation to 5 seconds, and therefore it is still many times smaller than when one looks at a star with the naked eye. Its apparent diameter is always at least 2 minutes and therefore 24 times greater than through the telescope. Its light must therefore appear

600 times weaker to the naked eye. And if I stay with $\frac{1}{2}$ second then the light is 60,000 times weaker.

[182] In respect to telescopes the various distances of fixed stars have nothing to say because those distances can just as well be considered infinitely great. On the other hand, the distance decreases the apparent magnitude of the stars. I will assume that a telescope is so perfect that it brings parallel rays to one point in the eye; then the image of distant stars will certainly become smaller but will always retain the same clearness. The whole matter turns therefore on how small a point on the retina has to be, until so strong a light as the one coming from the stars can no longer make any impression when it falls only on this point. Since the optic nerves can be so sensitive and small, this point always has to be immeasurably smaller, so that the light falling on it could not set the bundle of nerves into a perceptible agitation.

But in spite of the fact that neither the telescopes nor our eyes are so perfect that each geometrical point should appear as such a point on the retina, they nonetheless approach this perfection because the telescope can be adjusted to the eye. Thus a point can be recognized whose apparent diameter viewed through the telescope hardly amounts to a second. The telescope presents the situation as if we were seeing the image of the point at a distance of 8, 10, or 12 inches. At that distance we would clearly distinguish, [183] according to Mr. Musschenbroek's experiment, a silk thread $\frac{1}{1948}$ inch thick.[5] Its apparent diameter is less than a second. A small spark which is brightest as it is extinguished would appear to us as clearly if it were not thicker than such a silk thread.

Since the strength of the light of fixed stars when seen through the telescope is not changed by the various distances, but only its apparent magnitude, I can push a star back many thousands of times more, and it still would be seen through the telescope. Its light will still remain strong enough to produce an impression on the eye. On this point I make the following reflection.

When the image of a star on the retina is smaller than an optic nerve, it still puts that whole nerve into agitation, and since its image in the soul arises through that agitation, I will conclude from this that all stars must appear to us when viewed through the same telescope as equally big, namely, as pure shining points. I would, on the contrary, find a difference in their brilliance because the vibration of the nerve is the weaker, the farther is the star. It is easy to see that this vibration admits innumerable degrees before it becomes insensible. And since we experience [184] that vibration even at night, because even then we see objects, we can draw from this a conclusion about the degree of sensitivity of the optic nerves. We could still perceive a star if its light were many times weaker than an impression from a house or from a sheet of paper on which the moon shines. The light of the moon is 500,000 times weaker than that of the sun or of a fixed star,[6] and the brightness of the paper which is illuminated by the moon is hardly the 100,000th of the brightness of the moon. Also, the light of a fixed star could be weakened 50,000 million times and it still would shine as brightly as a paper which is illuminated by the moon. Now calculate, Sir, how far should one remove Sirius until its image when seen through a telescope would become as weak as the light of such a paper?

Thus the telescopes do not give us so much the service of removing the foreign and scattered light which makes the image of stars unclear. For in this case the greatest part of the light would necessarily fall away and a star would not appear at all sparkling. On the contrary, the telescopes unite all this light in a point, and all the more exactly, the more perfect they are. The point thereby retains all its brilliance, and this has necessarily to be all the more vivid, [185] the more sharply it is impressed on the retina.

We see this in the satellites of Jupiter and Saturn that remain invisible to the naked eye because their light is scattered too much. The telescopes collect it and bring the image of these satellites to almost a complete distinctness, so that we see them almost as bright as their planets. We would see them as completely bright if the telescopes were able to present them completely clearly.

You can conclude, Sir, from these reflections that I can without any hesitation consider as recognizable the fixed stars that are many thousands of times farther from us than those of the first magnitude. I readily grant you that the systems which I place in the Milky Way behind one another demand such distances. An easy calculation will evidently show this. Galileo has counted 400 stars between the Sword and Belt of Orion.[7] The area occupies about 10 square degrees.[8] The whole sky has, as does each spherical surface, 41,253 such degrees. Therefore counting 40 stars for each square degree, there will altogether be 1,650,120 stars if they were equally dense in all places. If I place now 12 stars next to our sun, 4 times 12 in the second, 9 times 12 in the third distance,[9] and I let the number for each successive distance grow as the squares, [186] then I find that, counting our sun, there ought to be such distances behind one another until the sum of all stars comes to 1,650,120. Accordingly, we would have in our system fixed stars of the 75th magnitude,[10] although we can see with the naked eye only stars of the sixth magnitude. If I now take the distance of Sirius from our sun as the yardstick, then I must apply it 150 times to measure the diameter of our system. If I put, according to my last letter, the nearest system in the Milky Way still 10 times farther, then I come to 1500 such yardsticks which consequently would do for about the inner diameter[11] of the Milky Way. The outer diameter could not be determined by so small a number because, in order to fill that streak so thickly with systems, I must place a good many hundreds of these behind one another.

At any rate, I do not think at all that our best telescopes are good enough to let us see clearly the outermost stars of the Milky Way.[12] They necessarily must, because of their immeasurable smallness, vanish gradually and form a fused, very weak light, especially if the light of fixed stars ought to weaken gradually in the ether, which is very probable after what I have noted in my last letter.

[187] The other question which you, Sir, have submitted concerns the gravity of planets toward the center of our system of fixed stars. You will see from my previous letter that I still leave open this question whether this gravity is considerable enough to cause in the orbit of planets observable anomalies, which should be very similar to those which we observe about the moon and which derive from the moon's gravity toward the sun. One should see whether the earth, for instance, moves in a simple ellipse, or whether it deviates from it annually as the moon does from one conjunction to another. I have incidentally raised the question whether the gradual motion of the aphelia and nodal lines in the planetary orbits might not be derived from the planets' gravity toward the center of the system of fixed stars.[13] Were this the case, then this gravity had to be very small, not because it was small in the center itself but because our sun's distance from it is very great. What matters here is whether the diameters of planetary orbits have significant relation to this center. For then it becomes clear that even the moon should have many more considerable anomalies if its orbit around the sun were greater, because these anomalies depend largely on the changes of its distance from the sun.

Furthermore, these anomalies would recur [188] in the paths of planets with each orbiting. They obey a double pattern. On the one hand, they ought to be greater in each orbiting with the superior planets, because their path and consequently the difference of their distance from the center of the system of fixed stars is greater. On the other hand, these anomalies recur in the case of the superior planets more slowly because these use more time for their orbiting. This brings the periods of the retrogradation of perihelia and nodal lines more to equality.

About the fact that the nodal lines of planets move in the direction of the sequence of the signs [of the zodiac], whereas those of the moon move backwards, one can make several remarks which can be considered as so many ways for resolving the doubt which you, Sir, have presented to me. Since we do not know yet the direction of the sun's orbiting, you can easily realize that it should suffice now if I show the possibilities of which there are several.

First, I have already remarked that the inclinations of planetary orbits are reduced to the earth's orbit only because astronomers can thereby facilitate their calculations. Exactly the same is done by astronomers on other planets. They look at the plane of their orbit as immovable [189] and they reduce to it the orbits of the planets. They have therefore quite other nodal lines than we do, and we are in agreement with them only in that the nodal lines which they give to our earth's orbit have the same position which we give to theirs, because the line of intersection is one and the same for both orbits. But the advance of this line is in the opposite direction when projected on the orbit of a planet or on that of our earth.

But this advance should not be considered in such a way; if, however, one wants to find its true advance, one has to project it as far as the fixed stars, and consequently to derive the precession of equinoxes from its yearly motion from the head of Aries.[14] If I take the motion of nodal lines as given by Kepler[15] and derive from each the motion of fixed stars which they appear to have around the pole of the ecliptic, then the nodal lines of Saturn and Mercury will move with respect to the fixed stars in the direction of the signs of the zodiac, but in the opposite direction in the case of the rest of the planets.[16]

This would be even more against a general law. But it is to be noted that one should not consider here the intersections which planetary orbits make with one another, but those which they all make with the plane in which the sun moves around the center [190] of the system of fixed stars, which, however, is still to be found. In general, it seems as if this plane had no marked inclination with respect to the ecliptic.[17] It may be but slightly different from the plane of the sun's equator. Still nothing can here be determined by pure speculations, because all circumstances must be taken together if one wants to trace out the center of the system of fixed stars, the plane in which the sun moves around it, the distance of this center and its gravity.

Before, however, one goes that far, is it not wisest to investigate first through careful observations the orbit of our earth whether it is in fact perfectly elliptical, or whether it deviates in various places, and whether the earth moves now faster, now slower than what is implied by Kepler's law? For if the earth had a considerable gravity toward the center of fixed stars, then it would move faster [at the point] where it is in conjunction and opposition[18] with the sun and with that center, as we see it with the moon when it is new or full. This would be the next method to determine approximately the place of that center which you, Sir, place for other considerations in the area of Orion, because it certainly appears that the rows of fixed stars are much longer in that area. Until a more exact determination [of that center] I do not hesitate to picture the system of fixed stars [191] in such a way as you present it at the end of your letter and to look for the center of our system and for that of the Milky Way in the area of Orion.[19] For the time being I do not see yet that here something more precise can be found, and I am only concerned lest one might consider this effort too daring. Still I would not be detained at this point. I know all too well that when one submits a hypothesis it is better if one elaborates it completely and presents it in its [full] extent. Since I afford the means of proving it, this is all one can demand, and after such probing it becomes clear what there remains for improvement. These means all rest on the question whether the sun has toward the center of the system of fixed stars such a gravity that its effects on the planets can be perceived, small as they may be. If the gravity is not so great, the means given are not sufficient, [and then] the question remains undecided, and the exact determination of the position of our sun and of its orbit will not be settled so easily.

For the time being I stay with reasons afforded by analogy upon which I have built the partition of fixed stars into special systems, although I cannot go beyond a certain degree of probability which everyone can determine according to the difference of his insight. So much [192] seems, however, clear enough to me that one cannot consider the Milky Way in a more fitting way than by dividing it into single systems of fixed stars and by turning it into a composite system. Its figure gives us the immediate lead to this. In those reflections which I made about it I find yet no contradiction. They rest on such propositions as can be sufficiently proven. The motion of fixed stars becomes just as necessary because of the law of gravity. The considerable distance that ought to exist among the nearest fixed stars, if they, like our sun, should be useful through their light and heat, demands that one should place them in long rows, one behind another, and as a result the whole system of the Milky Way necessarily becomes flat, because the rows of stars in that streak must certainly be many times longer. In general, I find not the slightest reason to part with those conclusions which gradually led me to such a conception.

Further I did not proceed in the matter, although I often had the desire to investigate how far the systems in the Milky Way might be distinguished from one another, because we indeed see the Milky Way divided here and there. Although I place so many systems in this streak behind one another, so that they form in our eyes a coherent whole, I still do not feel confident to determine [193] anything, and before I proceed further I wish that before one first accomplishes the projected proof through exact observations, one especially investigates to what extent the fixed stars might have been displaced with respect to one another since the time of Ptolemy. Afterwards, one could also investigate whether the motion of aphelia and of nodal lines in the planetary orbits might be deduced from the given reasons. Since this investigation can be done for its own sake, it may easily turn out that I have already proved something with it. But the work demands more time than I have. Meanwhile, it will always be a pleasure for me if you, Sir, would take the trouble of thinking over the matter even more and communicate to me your reflections on it. But it would be an even greater pleasure if your time and circumstances permitted you to come again to us. I hope with the greatest longing and rejoice in advance, because it will be most pleasant for me to show you that no devotion is more perfect than the one with which I remain, Sir, etc.

[194]
FIFTEENTH LETTER

Strength, much rather than desire, would ever fail me in my pursuit of the worthy encouragements which you, Sir, give in your precious letter to make each sky the object of my reflections and to carry myself in thought to the farthest of them out yonder. You may be assured that I wished nothing more than to become a Magellan[1] in astronomy and, even if it took many years to have the fastest of my thoughts complete the journey along the outermost periphery of the world-edifice, I would come back satisfied and enter into my diary larger numbers of fixed star systems and of Milky Ways than those which stand for the islands of the Indian and Pacific Ocean.

But my strength does not extend so far, and I will always be satisfied if I can follow you, Sir, in everything. I know that you are not pleased with giving free rein to imagination which after piling up millions of horizons gives nothing more than the magnitude of a space, which would only be an empty picture of the extent of the world or perhaps even of merely one part of it. You much rather would labor to pave the road as you extend it [195] through the depths of the firmament and to connect with conclusions thoughts which [otherwise] can easily fly away. Thus you always know how far you have advanced and you always find the return path when you pause to survey the road already traversed or to prepare yourself for further advance.

In particular how pleasing to me is that caution which makes you hasten slowly, so that I find the road paved and can follow you, Sir! You remove from my way the obstacles which would hold me up, while you take pains to clear up my questions. Distant as you place the stars in the Milky Way, you still lead their light into our regions and lift with elaborate reasoning the doubt which I had about their visibility. I see now that if this light becomes weaker this is to be ascribed not so much to our telescopes as to the ether and to the atmosphere of the fixed stars, and that the telescopes would present us each star in its true brightness if they could make its picture perfectly distinct. But since you bring this image very close to this distinctness and since the ether in all evidence must be very pure [transparent], I have no more objection with respect to the visibility of stars.

[196] Because of your considerations I am now equally free of doubts about the motion of the nodal lines of planetary orbits especially since you show several ways how the question raised can be explained, and you refer the decision to a more exact investigation as to which of these possibilities is really the case. For the time being it is enough that several cases are possible to remove the difficulty that occurred to me, and this serves for me as a new proof that even here we do not think yet in a sufficiently Copernican way if we want to make the plane of the earth's orbit the basis of the inclination of the remaining planetary orbits, because this happens just as much for the convenience of computations as when we assume in spherical astronomy that the whole firmament turns in 24 hours around the axis of the earth. I now see perfectly that as long as the nodal lines move, the earth's orbit is just as much excluded from the supposed immobility as are the orbits of the other planets.[2]

But since you look upon the more precise determination of this retrogradation [of nodal lines] as an opportunity through which one could demonstrate a posteriori as well the arrangement of our solar system and in particular the orbiting of the sun—which I would consider as sufficiently demonstrated from the general reflection which you present—it is therefore my true pleasure that I can share with you on this point [197] a piece of news which a friend of mine gave me. You remember, Sir, that you raised the question whether it would not be possible to demonstrate from the comparison of old and new catalogues that the fixed stars change their positions with respect to one another and that they gradually shift from their places. You can now consider this question clarified with respect to many stars. My friend told me that Prof. Mayer of Göttingen,[3] whom we have to thank for so many other discoveries, took pains to pursue further these investigations which he had already undertaken. He has sufficiently explored the matter and was led to the conclusion that one cannot doubt that one would find after a general investigation that all the fixed stars have shifted more or less.

You can imagine how welcome this news was to me and how eager I am at present to read the paper[4] which contains an evidence so important for all astronomy and for the whole determination of its system. Meanwhile, I took the observation with joy and mustered all my strength to see what I should conclude from it.

I began by taking the return road and here your previous letters were a welcome help. For you, Sir, [198] have derived this gradual displacement from the law of gravity, because this law adheres to each particle of matter and therefore it is not only spread across the whole world but especially serves the purpose of making the world a connected whole. This law seemed to you for good reason to demand also a centrifugal force, because the preservation of celestial bodies in their appropriate distance rests on it. Thus you have derived both central forces, and the displacement of fixed stars was a necessary consequence of them.

Henceforth I can turn this proof which is completely synthetical into an analytical one, because the last conclusion transforms it into an indubitable experimental evidence which now can be laid down for foundation.[5] I feel confident that I can derive from it the central forces in a necessarily conclusive manner and I want to submit this attempt to your judgment which I esteem very highly.

Since now it is an experimental evidence that the fixed stars move, this motion is necessarily either rectilinear or bends in a curve. The rectilinear motion borders on the absurd. For if it is convergent in all stars, then one must assume that the world was so created that it should gradually collapse into a chaos. This would be contrary to the [principle of] preservation and therefore absurd. [199] If the motion is divergent, then the whole universe gradually expands and its parts scatter into the infinite. The whole will shake, its common tie will fall apart, the order will vanish, in brief, the divergent motion becomes absurd.

I can add that the rectilinear motion would be too uniform to find a place in so complicated a machine. But this motion is already absurd enough to require its complete exclusion in connection with the celestial bodies. From the mechanical principle, that each moving body keeps its original direction if it is not deviated from its path by some new force,[6] it follows necessarily that the fixed stars will always [never] be deviated from the straight line. If the world is not to fall apart, then not only the central forces will be necessary but also such a balance of them which keeps the celestial bodies in well determined orbits. There must be present with each star a force which always brings it closer to the center, from which it would move away because of motion.

Henceforth I extend to the whole world the Newtonian law of gravity, [and] it may now either belong as a necessary consequence to the matter of which the celestial bodies are made, or it may only be introduced for the sake of a splendid harmony [200] which you, Sir, put in its true light in your letter. If gravity as an attractive force resides in the body itself, then the fixed stars are much too similar to our sun so as not to attract exactly as does the sun. If gravity depends on the pressure of the ether,[7] then I would spread it out as far as the light rays can go. And if both causes are at work, then again they are very similar in their effects, and the result must be one [and the same] because one similar law must bind each system of the world to one another.

In these reflections I have also done the following. Because the fixed stars are moving they have unequal distances from us. Otherwise, they should move on the surface of a sphere. This, however, becomes an absurdity, for you, Sir, have proved that the orbits of comets cannot also be great circles. Moreover, since each sun has a retinue of planets and comets which it illumines and heats, there would all the less be enough room on the surface of a sphere.

It will be a true pleasure for me if you, Sir, would take the trouble to pursue this conclusion further. For I am convinced in advance that from the displacement of fixed stars which you have [201] derived from general considerations you have drawn no further consequences only because these, according to your customary care, first had to be confirmed through experience. And for this reason you have submitted the matter only as a question whose clarification rested on the comparison of the old and new catalogues of fixed stars, just as you let it devolve to future, more exact observations [to see] whether the orbit of the earth and of the rest of the planets undergoes thereby a considerable change.[8] But since experience has already confirmed the first, I do not see what should hold up your inferences, and I beg you to communicate them to me.

While I am waiting for them with the greatest desire, I must still tell you that this observational evidence made me bold to search from every angle for reasons and considerations on behalf of your system. I wished to come gradually to your track and to familiarize myself fully with the art of concluding which you use here. For I already see that I would be put thereby in the position of foreseeing future observations and of waiting for them confidently as proofs of my conclusions. Although I am not yet in full mastery of this, I must nevertheless tell you, Sir, how I made my start, and I know that you will help me to carry on. It does not concern the systems of fixed stars, for I must leave this completely to you until I acquire more expertise. Therefore I return [202] to our solar system to explore its arrangement, if I can, in greater detail.[9] You will grant me that it is worth the effort because it will give us a closer and clearer image of every other solar system. Halley's Table, which fell in my hands again, gave me the prompting. And since it contains so many already analyzed observations, it will serve my reflections on behalf of the goal toward which I want to develop them.

Since I know sufficiently that the determination of the path of a comet depends on 6 data [parameters] which necessarily distinguish it from any other orbit,[10] I have distributed the comets in classes with respect to each of these items and wanted to search for the regularity according to which they were distributed in each class.

The first class, which concerns the distance of the perihelia, you have already studied, and you have laid it down that the number of comets increases as the square of the distance of their perihelia. You have furthermore remarked that since Halley did not select the comets, but took them as he could have them, his Table must give a sample of their distribution, [and] that this sample with respect to visibility is open to an objection because certainly those comets, whose circumstances make them most readily visible, [203] ought to be more numerous within such a small number. I saw equally well that their regularity in respect to the distance of the perihelia could not consistently be found in Halley's Table. However, it applies fairly accurately to comets which have their perihelia closer than Venus. The orbit of Venus is about 3 times greater than that of Mercury, and the Table gives 17 comets which have their perihelia within Venus and only 6 which have it within Mercury. Now 17 is to 6 approximately like 3 to 1 and therefore like the square of distance of both these planets.[11]

This example which you have given me leads me eventually to look for such regularities about the other determining data of comets. I moved therefore to the inclination of their orbit toward the ecliptic and wanted to see whether all angles of inclination are possible. I took this angle from 10 to 10 degrees [in units of ten degrees][12] and watched how many of the 21 comets, which the Table contains, belonged to each group and how many should belong to each group according to the calculation. The result, together with the difference, I listed in the following table:[13] [204]

Inclination of the orbit	Number of comets	Computed number	Difference
10°	2	2+	0
20	4	5	+1
30	6	7	+1
40	11	9+	−2
50	11	11	0
60	12	14	+2
70	15	16+	+1
80	19	19	0
90	21	21	0

Since the computed number was now greater, now smaller, I concluded that, if Halley's Table contained instead of 21 comets a few hundred, the difference would not be much more considerable and that consequently I could divide the comets in such a way that all angles of inclination were equally possible and that therefore they would occur in equal number.

I similarly distributed the position of nodal lines across the signs of the zodiac. This distribution did not seem to me so orderly with respect to the individual signs. In Cancer and in Leo there were none, but in Gemini 5 and in Virgo 3, in Libra none, again in Scorpio 2, and so I saw well that the signs following one another compensated each other, because in the first 4 signs there were 9, and in the first 8 signs there were 14.[14]

[205] The latitude of the perihelia, or the elevation above the ecliptic, must increase like the zones of the [celestial] sphere and consequently like the sine of latitude, if they are evenly distributed across the sphere, that is, have all possible positions, and this was borne out very regularly in Halley's Table.[15]

Latitude	Number of comets	Computed number
10°	4	4−
20	7	7+
30	11	10+
40	14	13+
50	14	16+
60	16	18−
70	19	20−
80	20	21−
90	21	21

But it was otherwise with the heliocentric longitude of the perihelia. I first assumed that they had to be distributed equally across all signs, but the 6 northerly signs of the zodiac had exactly twice as much as the 6 southern signs. In those there were 14, in these only 7.[16]

In respect to the months in which the comets were in their perihelia the opposite was true, inasmuch as the number of those that went through their perihelia in the winter months was twice as great as the number of those which had their closest approach to the sun in the six summer months. Of the former there were 16, of these only 8.[17]

[206] From this I should suspect that the visibility contributes something to this discrepancy. Since we see comets only when they come to their perihelion, and furthermore we can see them only at night, the number of those that are visible in winter must therefore be greater than the number of those that can be seen in summer, and in proportion to the length of nights. The nights are, however, with us [at our latitude] twice as long in winter as in summer, and therefore it is also twice as possible to see comets in winter than in summer. For this reason I assumed that the perihelia have in respect to their heliocentric longitude all possible positions in spite of the fact that their visibility follows another law.

Finally, there remained for me the direction of motion, to see how many comets move forward or backward. Although Halley did not show this in his Table, it still could be determined from his data because he indicated the position of the ascending node and also whether the perihelion was southern or northern. For when it is southern, the comet moves from its perihelion toward the ascending node. If, however, the perihelion is northern, the comet moves from this node toward the perihelion. From this I obtained the direction of motion, and after I had done the survey I found that of the 21 comets of the Table 11 were moving backward and 10 forward, and that consequently one could assume equally many [207] of both kinds.[18] This makes the uniformity which we see in the planets even more remarkable, since all planets have the same direction.

With the comets both directions are so completely possible that I found no circumstance that changed anything there. Among the backward moving there were 5 southern and 6 northern, among the forward moving there were 5 southern and as many northern, and both the forward and backward moving had all possible angles of inclination and perihelion distances.

From these considerations I would then make the general conclusion that the comets have all possible positions and varieties, and Halley's Table shows this in all kinds of scrutiny, but it still leaves unusually many empty places and gaps which still are to be filled if it is to become complete one day and contain all the comets visible to us.

I see it well that if these considerations are to be useful they should be applicable to a certain kind of [calculation of] probability which I cannot yet develop in full. I take, for example, as a basis that all possible positions occur in cometary orbits, and from this it can immediately be determined how the comets are distributed [208] in respect to each determining item [parameter]. All angles of inclination are equally possible. Each sign of the zodiac has an equal number of nodal lines and perihelia. The number of comets in respect to the heliocentric latitudes must increase as the sine of latitude, and therefore always slower, until it stops at the 90th degree of latitude and becomes complete. Finally, this number also increases as the square of the distance of the perihelion from the sun.

After these hypothetical determinations I turn to Halley's Table to see whether it follows them or shows [no] unobjectionable deviations from them. The inclination of orbits, their nodal lines, and the latitude of perihelia go as desired. But the distances of perihelia and their heliocentric longitudes as well as the months of the appearance deviate from the conclusion derived from the hypothesis, yet in such a way that this deviation can logically be explained by considering the visibility, if one transforms the cases possible in themselves into such that are more or less possible in reference to visibility. This transformation I would not yet determine precisely enough in reference to the distances of perihelia, because they seem to depend on many circumstances. For the time being it is enough for me that comets which almost necessarily ought to be visible follow that hypothesis, and it seems to be valid even for those comets that cannot be seen so easily.

[209] The determination of this probability which takes place here seems to be founded on that hypothesis. I must assume that there is no selection in Halley's Table. For this is precisely what makes the matter applicable to the theory of probability, just as it happens with throws of dice. Furthermore, I must assume the return of comets to obey such a very composite order that they follow one another in such a way as if all cases were possible, but that the case which was the most probable had the upper hand.

With such presuppositions I picture the matter in the following manner. I assume a general law and from this I determine six special laws for the determining items [parameters] of each case. Of these individual cases I take 21 exactly as they come and without selection, and I compare them with each of these special laws. They invariably hold true. If one were to take instead of these 21 cases all those that actually take place, there would arise the question how great is the probability that they do not obey at all those six laws which all follow from one general law.

Although I cannot yet answer this question and in addition it seems to demand extensive computation, I nevertheless perceive this much— that because Halley's Table obeys these laws [210] very regularly the comets missing in it, and therefore all comets, must obey this very same law, because it seems impossible that no notable exceptions should occur in this Table from most if not from all these laws, if the total number of all comets would follow a distribution of another kind. For how could it happen that these 21 comets appeared and were observed only to deceive us in all these six different respects?

It would pay, in my opinion, the effort to put this speculation in full light when the opportunity arises to determine the gaps of Halley's Table. But you, Sir, have already utilized in this connection the distance of perihelia, and [since] this [parameter] is the most suitable, I will merely undertake a gleaning after harvest when I will look for the solution from the remaining pieces [parameters]. The position of the longer axis of the orbits seems to me here to be of the best use. The two that come closest to one another in the Table[19] form an angle of about 7 degrees with one another. If I distribute them through the whole surface of the sphere, I must divide 41,253 square degrees, the amount [area] of the surface [of the sphere], by 49 [7 × 7] square degrees, and I get 842 comets visible to us which all come inside Mars,[20] and 40 times as many, or 33,680, which come inside Saturn.[21] The number would again be very [211] considerable. Still the computation is by far not as conclusive as the one which you have derived from the perihelia, because the perihelia imply in a more stringent manner the sphere of influence of a planet than the mere position of their longer axes.

What appears to me truly remarkable in all this is that all angles of inclination are equally possible, because I always thought that they should be considerably greater with comets than with planets. But Halley's Table contains such whose angles of inclination are not more than 5 to 6 degrees, and therefore not greater than that of Mercury which Kepler puts at 6 degrees and 54 minutes.[22] So closely therefore do the world of comets and the world of planets border on one another that they gradually fuse, and in all likelihood the angles of inclination of cometary orbits lying next to one another will not be much greater, because they still should evade one another just as well as do the planets. Such a small inclination seems, however, to show that the sphere of influence of comets should not be considerable, and Jupiter and Saturn would still retain the ranking position among the celestial globes visible to us. Once the comets luckily pass by these [two planets] I am not concerned with other eventualities.

Write to me again, Sir, how greatly the points, which I tried to bring up, [212] can serve you to deduce from them still further conclusions; but I particularly hope that you will communicate to me your reflections and inferences about the displacement of fixed stars, for which I will be extremely obliged. Consider this letter as a proof of the labors which I impose upon myself to continue on your road and to make mine your way of thinking in the astronomical mysteries as well as in the sentiments of true friendship, with which I remain, Sir, etc.[23]

[213]

SIXTEENTH LETTER

It was an exceptional satisfaction for me to see from your precious letter how you, Sir, do everything to gather new material for our astronomical reflections and to make these more certain not only in general respects but also more precise in each individual detail. I readily admit to you that I believed I was at the end of my insights, and I soon would have left it to future eventualities to go any further. So little can we rely on our thoughts when we are to find members of a long and interconnected series of conclusions! Soon we lose sight of the side of things conducive to new propositions, soon we fail to perceive what can be useful in them, and soon we are without the premises which we ought to attach to them to bring out new conclusions. Much of this depends on time and circumstances and makes us wait until the solution of questions offers itself on its own. Accidents are aplenty in our investigations and we ought to be very attentive to recognize the useful occasions and opportunities and apply them.

How much I wished to be able to utilize happily the news about the displacement of stars, now set beyond doubt, a news which you have communicated [214] to me, and to pursue further the conclusions which you, Sir, have already derived from them! You have all the reasons in the world to consider this experimental evidence as very important for the whole astronomy. Whatever in our solar system is subject to change, however slow, one has always tried to establish it by a comparison with the fixed stars until now. What the shore and capes are in navigation, the fixed stars were in astronomy; one imagined to find them always right there. But now that it is entirely proven that the fixed stars move, one has to think of other means to find again in future times the astronomical places which one observes. The moon which now covers a fixed star will not cover it in future centuries when back in the same position.

Although I considered my general reasons for the displacement of the fixed stars too satisfactory to permit any doubt, I rather doubted whether this displacement can already be observed. Therefore I proposed the comparison of old and new catalogues. But because Prof. Mayer[1] found that displacement noticeable, it is unquestionable that I can now follow his proposition and see what is there for drawing further conclusions.

[215] So much I can foresee: that from such a comparison not so much the magnitude as the reality of this displacement can be exactly established. For this displacement must be notably greater than what can be ascribed to the faulty observations of ancient astronomers alone, and this in itself is already enough [for the purpose]. The astronomical instruments are today so perfect that one will be able to determine this displacement in much shorter time than this could happen from the comparison of old and new catalogues [of stars].

Once, however, it is determined, one can certainly make the calculations which you, Sir, mentioned in one of your previous letters as an incidental project. You assumed, for instance, that the nearest star moved by $\frac{1}{4}$ degree and from this you have calculated the velocity of its motion and concluded that it cannot be caused by its gravity toward our sun. This is reason enough to look for another center for our system of fixed stars, and all the more so because the displacement in all evidence will be greater than $\frac{1}{4}$ degree which I let stand for the time being until Prof. Mayer's paper becomes available.[2]

Then you have, Sir, already remarked in the same letter that this displacement [216] is only relative. For you can rightly assume that our sun remains just as little in the same position as do the fixed stars, [and] therefore the change of the sun's position becomes mixed up with the apparent change of the stars' position, and it is undeniable that one here has to distinguish the optical from the physical as this has already been done for a very long time in the whole astronomy.

But to put this matter in its true light, I must tell you that I am now no longer concerned with the determination of the distance of the fixed stars from one another. You know how much effort was given to using the parallax of the earth's orbit for finding the distance of only one of the nearest ones, and that the radius of this orbit hardly bears a relation [proportion] to that distance because the latter is at least 500,000 times greater.[3] But now we have quite a different parallax which will gradually give the position of the fixed stars more exactly. It is the parallax of the sun's orbit which the sun traverses around the center of the system of the fixed stars, and we now can use in relation to the fixed stars the same devices of which the astronomers on the moon had to avail themselves in respect to the planets if they could never see the sun. It is natural that the same process will cost more time for us. The magnitude and worthiness of the cause demand it.

[217] One can see the portion of the orbit which our sun traverses in a few hundred years as a base line, and it is on this that the displacement of the fixed stars should be projected. We find ourselves in a similar situation when we want to find the orbit of a comet from three observations. The difference which increases the difficulty is that we still do not know the center of the system of fixed stars. For it is to this center that one must relate each angle.

In my previous letter I have proposed different means for finding that point and let it be reserved for further investigations to ascertain whether they are satisfactory or not. Since the apparent displacement of the fixed stars depends as much on the motion of the sun as on their own apparent motion, one may perhaps derive from this in which direction the sun is moving. It is unquestionable, as you, Sir, have already remarked in your previous letter, that some fixed stars must appear to move forward, others backward and still others as standing still. The motion of our sun comes into play in all and produces an optical irregularity from which the true motion must be recognized though very gradually.

Moreover, I picture to myself the system of fixed stars to which our sun belongs not otherwise than [218] our solar system. This is also a reduced image of the other. It is probable almost to certainty that the orbits of fixed stars can have not only all possible positions with respect to one another but also can be ellipses of all kinds. The first is clear from the fact that we see the fixed stars distributed everywhere around us, and this could not be so if they lay in the same plane as the planets. The other seems to me to be demanded, as in the case of comets, by the variety and increase of their number. For you will, Sir, still remember that the elongated ellipses are more suited for the increase of their number. This I again make necessary from the notion of habitability. Each fixed star has a retinue of comets and planets. If these are numerous, I also must multiply the orbits and bring into the system of fixed stars the greatest possible number which it can take. It must therefore be clarified from future observations whether our sun follows a more elongated or a more round ellipse, and whether its orbit is more or less inclined with respect to the plane of the Milky Way.

The true velocity of the motion of fixed stars is unquestionably most varied, and the smaller it is, the farther they are away from the center of the system. I have already remarked that our sun ought to be rather close to the center and this makes the velocity of its motion [219] greater. On the other hand, there may be some fixed stars of the sixth magnitude that move much slower. Among these I count those that seem to stand opposite to the center, that is, they are in opposition to it.[4] This would again afford a means for determining eventually the area of the system's center.

These are more or less the reflections that occurred to me about the displacement of the fixed stars, a question which now has been settled. You can easily see that all these reflections are aimed at [the project of] turning these observations in the future to good advantage, and because of this I have set them forth in a more elaborate manner than if I thought that one could find something more definite in a short time. Thus I could display methods somewhat useful for that purpose and find my satisfaction with what may be useful for the confirmation of the theory about the distribution of fixed stars in special systems, and you have, Sir, already given the reasons for this. They will depend on these points: The fixed stars move in orbits. The central forces are at play there. The fixed stars gravitate not toward the sun but together with the sun toward the center of the system and move around it in orbits which allow all possible positions with respect to one another.

Do you wish now to give free rein [220] to your imagination and see whether there are hyperbolas as well among these orbits? Then you will find suns which continue their course from system to system or from Milky Way to Milky Way. Here number and measure will fail us and it will become clear from everything that the world is not created for the sake of a moment. The question is whether a Platonic Year[5] is sufficient for our sun to go around once in its orbit, or whether it would traverse in such a year but a sign of its zodiac. Since the sun is near the center of the system, this [Platonic] year still might be very small compared with those which are needed by the outermost suns of the system for completing their orbits. What will the year look like during which a whole system moves around, and in how long a time do we want to lead the Milky Way around in a circle? Shall we call such periods the moments of eternity? They will soon begin to have some relation to it.

I fully agree with you, Sir, when you say that our solar system is an image of every other [solar system], because similar means presuppose similar purposes.[6] I would go here still further and state that our solar system is also an image of a system of fixed stars inasmuch as the distribution and orbits of the fixed stars have a very similar arrangement and in all of them some central forces are introduced to maintain the stars in definite [221] orbits. The only point where I still cannot make a decision is whether I should leave empty the common center of the system, or place there an enormously great dark body whose mass should be heavy enough to keep each fixed star in its orbit just as our sun does this with respect to the planets and comets. We still cannot say how dense that body should be if it no longer contained intervening spaces [among its smallest parts]. Perhaps gold, the heaviest of terrestrial substances, ought to be considered a sponge compared with such a dark body.[7] I therefore did not concern myself about the size and mass of a body which could direct around itself a whole system of fixed stars. But I still find more reason to provide it with all light through letting it be illumined by the nearest star which orbits around it. Were it itself bright, each sun around it across a vast space would be superfluous, because I consider the light only as a means serving a dark body.

If, however, I assume that the center of the system of fixed stars is quite empty, then these have no other gravity than the one which each has toward all the rest. Those closer to the center of the system will gravitate toward the outer ones and therefore in opposite directions, and I do not see how a central force and a uniform motion would result. The mean direction [222] would in the case of each star be composed of millions of directions and therefore changeable in each moment. The regularity of paths would be much too complicated to lend itself to a system whose size and duration demand something very simple. How little and how rarely do the planets disturb one another and how much is their motion made simple by the powerful influence of the sun! We see these smaller displacements as exceptions that perhaps also have their usefulness, though regardless of this they remain exceptions because they are deviations from a general law. Accordingly, if I leave empty the center of a system of fixed stars, it appears to me that I am taking away everything general in the laws of their motion and leave behind nothing but such exceptions. The order then has to be looked upon as a constant mixture of millions of individual influences in much the same way as this happens in the course of things on our earth. But I do not consider in such a fashion the arrangement of the world-edifice. The closer a system comes to the whole, the simpler the laws of its changes ought to be.

Add to this, Sir, still the fact that because the nearest fixed stars are so far removed from one another, they influence one another not only by themselves but together with their whole retinue. In your last letter you have again submitted an estimate [223] of the number of comets. I do not know whether the five millions which I first reckoned might not seem too many for you. I do not wish to press the matter too much. But as you always realize it, in order to fill the gaps in Halley's Table you will still come up with plenty of comets. I assume without hesitation only as many of them as can orbit without disturbing one another. You are ready to grant that much, and thus the mass of planets and comets taken together becomes much greater than the mass of the sun.[8] All the more considerable therefore will be the gravitation of a whole solar system toward another, and all the more one has to be concerned lest they disturb one another more than the system of Saturn is disturbed by the system of Jupiter. How will you implement this necessary and healthful purpose without filling the center of the system of fixed stars with a body of enormous mass?

If, however, such a body is in the center of a system of fixed stars, such a system will become wholly similar in its arrangement to our solar system, and the difference will only consist in that in the latter dark globes orbit around a bright one, and in the former bright ones with their retinue around a dark one.

Once this similarity of arrangement is presupposed, [224] the reflections which you, Sir, have made about our solar system are very useful in making us more closely acquainted with both [the solar system and the systems of fixed stars]. You fully compensate with those considerations for what I have overlooked in my foregoing remarks about Halley's Table. It can be explored according to these six viewpoints that are taken from the data [parameters] determining each orbit, and I have investigated the Table only from these viewpoints and rather incompletely. So much I see now that if one is to fill the gaps of the Table, the reasons for this would most conveniently be taken from the distances of perihelia. Still I soon saw as well that here various imperfections come into play which derive from a comet's visibility.

The possibility of seeing a comet on our earth depends in respect to the distance of its perihelion on various circumstances. Most important is its smallest distance from the sun and the earth which it has when it is above our horizon at night. If it is too far away from the sun, one does not see it because of the weak light. If it stays too far from the earth, it vanishes because of the much too small apparent magnitude. Both circumstances must co-operate if the comet is to be seen clearly and long enough, so that one may observe it several times and thereby determine its orbit.

[225] The computation which one could make on this point would certainly become very extensive and seems to me not really necessary, because it is enough to have here the general considerations, the like of which I have already presented in one of my previous letters. I then concluded that all comets must be visible to us if they come closer to the

sun than Venus, and up to that point their number in Halley's Table increases as the square of the distance. If they remain farther from the sun, the circumstances of their visibility become more and more restricted, the time of their sojourn in our sphere of visibility is shorter, and the position of the earth in its orbit is more definite.[9] Judging from Halley's Table as well as from observational evidence, these obstacles come very quickly together, because among the 21 comets there were only 4 which remained farther from the sun than Venus.

Meanwhile, it is enough for me that the remaining 17, which must necessarily be visible, agree with the law of the square of distance. Since this law is found correct from the sun to Venus, I see no reason why it should not as well extend up to Saturn and even farther out, although as a proof [of this] I know nothing but teleological reasons which I have already expounded sufficiently.

[226] The remaining viewpoints, according to which you, Sir, have gone through Halley's Table, are not subject to so many difficulties because of visibility. You have sufficiently developed the general laws for each determining datum [parameter] and I confide to you that I would rather have suspected that Halley's Table would more considerably deviate from them because it contains only 21 comets. You also strengthen the general law that their orbits have all possible positions [planes] in such a manner that each doubt about it becomes necessarily irrelevant. I thought it unnecessary to compute the improbability which you have pointed out in this connection, because it strikes the eye necessarily and very strongly once the question is posed as you, Sir, do it. For if you assume that the real distribution of all comets follows another law, then it is possible that the 21 comets in Halley's Table deviate from it more or less, but the deviation would be quite disorderly and one would necessarily have to choose these 21 in such a way that they should follow six special laws very different from the true one. I therefore consider it as practically proven that your observations on Halley's Table will be found all the more correct, the more it is brought gradually to perfection.

But that here a calculation of probability presents itself is certainly unquestionable, [227] and the two reasons which you, Sir, gave for it seem to me satisfactory. One can always use the probability there where chance events follow one another according to a very complicated regularity. The actual experience, when presented at length, always deviates somewhat from the highest degree of probability and is now too much, now too little, but never appears such as to become unrecognizable or that it should follow quite different laws. It always leaves very clear traces of the laws according to which chance events follow one another. Take, for instance, the mortality lists of a city from various years. If they are each year completely similar, one would then necessarily infer some necessity. But chance events always allow small differences though never such that

154

in one year there would be no deaths, whereas in another there would be twice or several times more than allowed by the number of inhabitants. In one year more lethal circumstances can accumulate than in another, but never so that all proportion to the rest should vanish. All causes that contribute to it run their course too closely to one another to be absent altogether at the same time or to be all together at another time.

It is true that the return of comets does not depend on so many causes. But their periods are incommensurable,[10] and the position which they should have at a given moment, for instance, at that of [228] the creation, is determined from innumerable viewpoints which depend on the most skillful arrangement of the world-edifice and on all its changes. Assume only, for instance, that the place and course of all celestial bodies are so arranged that they can in spite of the constant attraction always evade one another; then you will find that all chance here should necessarily be excluded, but that the situation is as entangled as if it were a mere happenstance. Thus the application of probability is appropriate and is really so in all individual events in which general laws are at work nevertheless. These always have the upper hand, and one will always demonstrate them from the observation of many individual instances in spite of their apparent disorder. Or when they let themselves be derived, as in the case of comets, from general considerations, this comparison can be found the more exactly, the greater is the supply of factual observations.

You, Sir, certainly have reason for pausing at the uniform possibility of the angle of inclination of cometary orbits with respect to the ecliptic, and for looking at the confirmation of this from Halley's Table as significant. Thereby you have also given me the occasion to think over the matter still more, and I readily admit that I cannot quite see it yet. The smaller [229] angles of inclination are presented in the Table as observational data and, instead of calling them in doubt, we should rather take them as a basis in order to infer, as you, Sir, have already done, the small sphere of influence of the comets. It is clear in itself that this conclusion rests on [the requirement] that the celestial bodies should disturb one another as little as possible because these exceptions ought to be very small.

Although I also accept the smaller angles of inclination because they are evident from observation, I have here quite a different problem. If I assume that the orbits of comets have all possible positions [inclinations], it seems to follow that not all angles of inclination are equally possible, but that the greater angles should be more numerous than the smaller ones. Here is my proof.

The position of a plane can be reduced to the position of a single line which is perpendicular to the plane. Since the plane of each cometary orbit goes through the sun, we can draw from the sun instead of a plane a line perpendicular to it, and extend this line to the fixed stars. The place where it reaches them is a point that can be marked on the celestial sphere, and we can call that point the pole of the cometary orbit just as our terrestrial orbit has [for such a point] the pole of the ecliptic. In such a way we can [230] omit the plane and the line because the pole of each cometary orbit completely determines its position.

If you assume now that the orbits of comets have all possible positions with respect to one another, then it is necessary that their poles should be uniformly distributed over the whole surface of the celestial sphere just as you have distributed the perihelia. As distant as the pole of a cometary orbit is from the pole of the ecliptic, as great also is the angle of inclination of these two planes to one another, and the complement of this angle to 90° is the distance of the pole of the cometary orbit from the ecliptic. Now you will find that as with the perihelia, the number of poles must increase as the sine of this complementary angle, and therefore as the cosine of the angle of inclination. This seems to me to be the law which is followed by the angle of inclination or by its complement. Take now the complements in units of 10 degrees and their sines so that the total sine be equal to the number of comets of the Table, or to 21, and you will find the following tabulation:[11] [231]

Compl. of angle of inclination	Number of comets	Computed number	Difference
10	2	4−	+2
20	6	7+	+1
30	9	10+	+1
40	10	13+	+3
50	10	16−	+6
60	15	18−	+3
70	17	20−	+3
80	19	21−	+2
90	21	21	0

It is obvious[12] that the difference between the observed and computed numbers not only is much greater than in the tabulation which you have given but that the discrepancy fell in all cases on one side. All computed numbers are greater than the observed ones but they were in your tabulation now greater now smaller.

What do you want to conclude from this? It is unquestionable that when one wanted to deduce a posteriori the law of the position of cometary orbits from Halley's Table, one had to fall back on the notion that all angles of inclination are equally possible. If, however, I assume that the cometary orbits have all possible positions, then quite a different law obtains, namely, that there ought to be more greater than smaller angles [232] of inclination. This, however, agrees rather poorly with Halley's Table.

I do not believe that the visibility can change anything here. For then I had to assume that the comets are much more visible when they have a small angle of inclination toward the ecliptic than when this angle is greater. This would do if the best conditions of visibility occurred in the southern hemisphere, because then the comet remains for us under the horizon.

Since you, Sir, have distributed the perihelia over the celestial sphere, as I did with the poles of cometary orbits, and since your calculation agrees completely with the Table, I doubted whether one did not stand in the way of the other. If such were the case, then the perihelia had unquestionably the preference, and the exception [objection] has to refer to the distribution of poles. If, however, both [calculations] were correct and the visibility caused no exception, then one certainly had to conclude that the ecliptic, or the plane of planetary orbits had something very special with respect to the angle of inclination. Perhaps one has to wait until Halley's Table becomes noticeably more complete, which is to be desired from various other viewpoints as well.

[233] It would be a real joy for me, Sir, if you had a happy insight for the solution of this somewhat complicated problem. I hope and expect it, just as the further comments which may occur to you about the still very unripe ideas of this letter, and with which you will extremely oblige the one who has the honor to be Yours with true devotion, Sir, etc.

[234]

SEVENTEENTH LETTER

Did you, Sir, indeed seriously believe that you have reached the end of your good and useful insights about the arrangement of the world-edifice? That I will never be able to imagine and your most appreciated letter is certainly no proof of it. It was for me [a cause for] modesty and new encouragements that you wanted to avail yourself of my cursory computations concerning the velocity of the fixed stars and of some of my other speculations, and to make them the starting point of your further conclusions. You have shown me in a most obliging manner how I could further develop my own thoughts and you gave me a welcome clue, for you showed me how these very same thoughts might have led me to the chief issue which I still did not see. Since for the time being I wish nothing so much as to take slowly a few secure steps and to familiarize myself

157

more with the means and ways which one has to apply to the firmament, it makes me rejoice that this first effort deserved your approval and that you are willing to show me its sequel.

I now understand fully that the principal issue about the recently discovered displacement of the fixed stars concerns the parallax of the orbit of the sun, and [235] to picture this for myself more vividly I have explored astronomy as it would be cultivated on the moon if the sun were invisible. To make the matter completely similar, I had to assume that the astronomers of the moon already knew that the moon orbited around the earth and the earth around some center around which also the other planets moved. I could further assume that they were familiar with the law of gravity, with the smaller anomalies that arise in the motion of the moon, with the indispensability of conic sections, and with Kepler's laws. I had to derive from these items how they would go about finding in a short time and in an approximate manner the place of the common center of [the orbits of] the planets, that is, the sun invisible to them. For that purpose the course of the moon and of the other planets could serve them. I can assume that the ellipse which the moon, without the co-operation [influence] of the sun, would describe around the earth is known to them, and thus they had to look for the places where the deviation which is caused by the sun, for example, in respect to velocity manifests itself most strongly. These were the places where the moon was shortly before in conjunction with, or in opposition to, the sun, and thus the position of the sun would eventually be determined.

The other method was to be taken from the planets. The astronomers of the moon had to observe their [236] position several times, and with the help of Kepler's laws their true place could be found in much the same way, though somewhat more laboriously, as we do this with the comets. The distance of the moon from the earth and its orbiting around the earth would give them a monthly parallax which they could use more conveniently than we can use the parallax of the earth's orbit with respect to the fixed stars. Furthermore this latter parallax would give them with respect to the planets and comets the service which we expect from the parallax of the sun's orbit.

The method which we have to use with respect to the position of the fixed stars is quite similar. The common center of the system of fixed stars is invisible to us, therefore its position must be derived through inferences which we cannot very well make unless the earth in its yearly course is subject to deviations, or we derive it from the small optical inequality of the apparent displacement of the fixed stars.

Although this is not going to happen soon, we can come ever closer to this aim, and gradually the true yardstick for measuring the distance of fixed stars will be found, a yardstick which is the distance of our sun from the center of the system of fixed stars to which it belongs. Our sun is not in that center. It will, however, [237] hardly be as far from that center as are some fixed stars.

Although here time and patience will be needed, it still satisfies me that there is also a method of knowing more reliably the arrangement of the system of fixed stars, and perhaps the posterity [in question], which I would put into the fiftieth century,[1] is not too removed from our times. Before I knew your system, as you, Sir, remember, I encouraged boldness on your part because I was not concerned about the judgment of so late a posterity. But your system needed no boldness, and if unexpectedly we ourselves were that posterity I would not be concerned that this judgment would undergo essential changes. It will require only a more particular determination which one has to obtain from observational evidence. You have provided well for the general principles in that arrangement. I now fully perceive the similarity between it and our solar system.

I still must devote some attention to the dark body which you, Sir, want to put in the center of the system of fixed stars and about which, while you find some problem, you offer proofs which one cannot pass over inattentively. Do you know what detained me? Most of all the consequences of your [238] proofs. You make those proofs so sharp, I believe, that they can be extended with the same sharpness to the center of the whole Milky Way and consequently to that of all systems of fixed stars taken together. You therefore prescribe an orbit to the whole Milky Way, because you ask me the length of time in which we want to lead it around. And the similarity which goes from system to system, the law of gravity which connects everything together and makes the world into an interconnected whole, the harmony and perfection which holds time and space together and therefore requires motion; in short, all reasons, from which you concluded to the motion of individual fixed stars and are now confirmed by experimental evidence, do not allow the Milky Way to remain at rest.

If, however, it has an orbit, your reasons about the dark body in the center of the system of fixed stars can just as well be applied to it. If you leave the center of the Milky Way empty, its motion will solely consist of small exceptions, because its particular systems have no common and dominating direction of gravity to make its motion as uniform as required in great world-systems. Each system of fixed stars comprises millions of suns and equally as many planets and comets around each sun. If you take all these bodies together, how considerable their common [239] gravity must be and how irregular their course when there is nothing that would force them all to remain in a very simple orbit!

I believe that I have proceeded in an orderly fashion in applying your principles. If I now take them as necessarily conclusive, I must then assume [the existence of] a dark body which has enough mass to keep the Milky Way in a simple order, and if you assume, Sir, that the Milky Way belongs to still innumerable other Milky Ways, then you give me material for a still many times heavier and larger body which must give law and order to all Milky Ways. However far you wish to go, you will finally come to the center of the whole world-edifice, and here I find my last body which steers around itself the whole creation.

Here I find material for my forces of imagination, and I count the moments of eternity in which the outermost boundaries of creation complete their circles. There is the throne to which all systems attend like so many satellites, the capital city that issues laws to the realm of reality and keeps all in order and complete harmony, makes all a whole, bans all excess, sets a limit to the revolt and dissolution of each fleeting part, and guides it back into its proper place.[2]

[240] But I unwittingly obscure by this enthrallment my difficulty which is caused by the view of such a splendid place. A poet would have here a touching opportunity to embellish it in full, depict it to our imagination to the highest degree of plausibility, and render it captivating. But speaking philosophically, I must admit that the enormous size of this dark body completely blocks my way. It strikes me as unbelievable. It is true that it may perhaps shrink if we assume that at least its core is of absolute density, provided that gravity is necessarily proportional to mass, or that the density of the ether increases in the direction of the center of the world-edifice and equally so toward the center of each larger system [of stars]. From what we perceive in our terrestrial bodies, the former case seems to be true because we find that gravity is proportional to mass, and so it still seems that one must ascribe to this dark body an astonishingly great volume.

Much as I am puzzled by this, I still cannot deny that your proofs appeared to me more enlightening on closer inspection. I have tested them on a small scale. I took the sun out of our system and therefore also the force pulling so strongly toward the center. But by the same stroke I also had [241] to decrease the centrifugal force, or the velocity of each planet and comet, if I wanted to make sure that they would not recede each and all from one another in a tangential direction and have the whole system scattered toward the fixed stars. But with that I have not straightened out anything. For Jupiter, whom the philosophers presented as a robber of planets,[3] would not really be such. Each comet in its vicinity would begin to describe ellipses around it. Its realm would become more and more considerable. The planets would themselves surrender to it, and even Saturn with its satellites would not be able to resist. Initially, things would go on in a disorderly fashion as in an uprising.[4] The sun has already been removed. Jupiter would see itself surrounded from all sides with globes.

160

The common center of gravity of all these globes, which at the start was very distant from Jupiter and kept changing its position, was now coming ever closer to it and finally coincided with Jupiter's own center. The gravity of globes toward one another remained without effect, because the gravity of those nearer [to one another] was offset by the opposite directions so as to almost vanish, and the gravity of those farther away [from one another] was united with their gravity toward Jupiter. Thus the system came into a steady condition under one smaller regent.[5] If now I take away Jupiter too, I believe that Saturn would next have the right to rule, and it would come about in a similar fashion. But through [242] such changes the [solar] system would become much less impressive, and the golden age would cease with [the removal of] the sun.

Judge now, Sir, whether I have drawn this picture correctly and to what extent it can be applied to a system of fixed stars. I believe I can conclude that without letting that dark body, the mightiest among the suns of the system, be in the center [and] appropriate supremacy to itself, the whole system would become much less significant in size, velocity, and simple order. So much is certain that under the rule of Jupiter, Saturn would be a mighty vassal for which a considerable number of smaller princes[6] would keep balance, so that it would not be subject to any perturbance. This, however, cannot be reconciled with the orderliness of orbits, for by giving overweight to Saturn at the point where it is, there ought to be opposite to it an even greater number of comets and planets which would pull Jupiter just as powerfully as Jupiter would be pulled by Saturn in the opposite direction. This [addition of extra retinue] is not at all necessary in the case of our sun, and the overweight which Jupiter and Saturn together give to one side of the system will be completely unnoticeable compared with the strong force of the sun and of so many comets around it. It seems to me that the common center of the whole solar system cannot [243] under ordinary circumstances be removed, perhaps even by an inch, from the center of the sun.

As little as I would leave it to Jupiter or Saturn to govern our system, just so clumsy does it seem to me that a system of fixed stars be governed by a fixed star. As a fixed star it must have its own retinue of comets and planets, but these were its very household; and I believe that it would have too much to do with them to be able to govern a host of fixed stars that are also its company. The supremacy must be all the more despotic, the more powerful it has to be. I therefore fall back on the idea of a body which can hold each fixed star with its whole retinue in harness.

That this body should be dark though illuminated by a sun, is very agreeable to me. The only thing that still gives me some concern is this: were such a body to be of an enormous size and illumined, then it always appears to me that some traces of it should be found,[7] since our sun is not at the outermost edge of the system but closer to the center and therefore closer to that body. It could therefore have a considerable apparent diameter, and its light, weak as it is, would not therefore be weakened to the extent that we could not discover it with our telescopes. Since the sun, which illuminates it, [244] orbits around it at a close range, it must have phases and should therefore now be full, now completely dark. If it had spots as the other globes, then even its apparent figure would change. Do you believe, Sir, that we may have some hope of discovering something of this kind? I wish that this would happen because the application of our lunar astronomy would thereby be greatly facilitated. It would no longer be necessary to assume that astronomers on the moon should first search for the position of the sun through circuitous routes for the establishment of their Copernican system. Since I am reserving judgment on this, I won't bother about what should be done in that case in order to determine more and more the position of the fixed stars of our system. However, I cannot deny that it would please me very much if I could soon find the means and method for that.

The comments which you, Sir, have made about my arrangement of the system of comets are all the more welcome, because I can consider them as the first try which I made to have myself acquainted with your way of reasoning. This much I see now that I succeeded in five out of six items in obtaining your approval, and especially that you not only approve my considerations about the computation of probability that arises there, [a fact] I was very pleased with, but you also [245] shed light on them through several considerations. But in connection with the sixth item you leave me in doubt; perhaps instead of telling me flatly that I have been in error there, you prefer to give me some useful exercises, inasmuch as you present to me the matter as a problem with whose solution I could practice myself. I love and seek the truth, and as you present it to me you oblige me. I will take it as such if thereby I can bring about the solution without being concerned whether I am to contradict myself and must change again what I have previously thought.

The difficulty which you present to me consists in this. I have considered in cometary orbits all angles of inclination as equally possible because they appeared to me as determining items [parameters] independent from the others, and Halley's Table prompted me even more to do so, however much I would have rather completely excluded the smaller angles of inclination. But the Table forbade this because it contains examples to the contrary. And this was precisely where I paused in my last letter and where you showed me how the matter can be considered from another side.

As to the conclusions which you, Sir, make in this connection I have to criticize them all the less because they flow in a much more necessary manner from a general law taken for foundation, [namely], that the cometary [246] orbits have all possible positions. You reduce the plane of each orbit to a line, and finally even more concisely to a point of the sphere, and these points must be evenly distributed on the surface of the sphere. It is clear that one speaks here of a physical equality. From this you derive that there ought to be more great than small angles of inclination. I thought it preferable to have this conclusion related to the ecliptic, although it could have just as well been referred to any other plane. For the time being I am still more preoccupied with the planets and with our earth than with the comets which adapt more skillfully to all perturbations and endure better all calamities.

Then you compare your computations with Halley's Table with which it should systematically agree, but in fact deviates from it very considerably and always on the same side, while my computation agrees with it systematically. The difference between the Table and our two laws is almost identical with the difference between the laws themselves.

It is unquestionable that if matters remained what they are I would be right, and on that basis you, Sir, have already offered the conclusion which you would draw: the ecliptic must have something peculiar to it. Only you wonder what that peculiarity may consist in. It [247] may very well be as if, in defiance of the planets, more comets had smaller angles of inclination than would naturally be necessary. What do you want to make of this? Are some comets and planets such good friends, or have Jupiter and Saturn pulled various comets so close to themselves that now the orbits are not inclined so steeply any more?

You may already see, Sir, that I am just about to start again with my lamentations in this connection because you still look for other means to remove this difficulty. You see the perihelia as distributed equally on the sphere—this is also the way Halley's Table presents them—and you ask whether this distribution might not stand in the way of the distribution of the poles of cometary orbits. I am unable to find here a problem at all. For, where I put one of these points, there remains for the others on the surface of the sphere a whole circle on which I can place them, whenever this is most convenient and, according to the even distribution, necessary.

I therefore hold myself to the other reason which you, Sir, give, namely, that the visibility stands more in the way of those comets which have a greater angle of inclination, because they can remain much more easily under the horizon during their best circumstances [of visibility]. You already see that this overthrows my rule, but I sacrifice it to the truth [248] all the more readily because I can thus leave the planets more at peace than if I had to assume more smaller angles of inclination. I have therefore sought whether Halley's Table might not contain a few traces of that obstacle which derives from the visibility.

To that end I assumed that the best conditions of visibility of a comet are tied to that part of its orbit which lies between both its nodes and stretches around the sun or to the side of the perihelion. This is brought out in most cases, although a few exceptions may occur.

From this I inferred that when the perihelion is southern, the comet in its best conditions of visibility can remain all the more readily under our horizon, the greater the angle of inclination of its orbit is to the ecliptic. However, if the perihelion falls into the northern hemisphere, the magnitude of the angle of inclination does not hinder the visibility; the comet will, on the contrary, rise all the more above our horizon.

I divided the comets accordingly in these two classes and found for the northern ones the angles of inclination 18, 31, 32, 32, 32, 55, 64, 79, 83, 83; whereas for the southern ones the angles 5, 6, 11, 21, 29, 37, 60, 65, 74, 79. The minutes and seconds I omit for brevity's sake. From this I saw [249] well that the angles of inclination of the southern ones were noticeably smaller, whereas with the northern ones the smallest amounted to 18 degrees. With the former the greatest was 79°, but with the latter two amounted to as much as 83 degrees.

In this distribution there are fewer comets in each class and therefore more gaps, so that I now compared them not from 10 to 10 degrees but only from 0 to 60 and from 60 to 90°. For, according to the rule which you, Sir, have given, equally many had to be found in both these intervals. Among the 11 northern ones 6 are between 0 and 60 degrees, and 5 from 60 to 90 degrees. Among the southern ones 6 are between 0 and 60 degrees, and only 4 between 60 to 90 degrees because with these the greater angles of inclination occur more rarely.[8]

I therefore do not hesitate to give up my rule which was to me objectionable anyhow, and the reason why that rule still fairly matched the Table is to be sought in that the visibility conflicts with the greater angle of inclination when the comets are southern. In such a way the ecliptic has no special importance, but the number of smaller angles of inclination becomes fewer, and this seems to me to be more useful also for the sake of mutual avoidance. The unexpected here is that the various degrees of visibility deviate so much from the real distribution of the poles of cometary orbits that [250] instead of your rule mine fits in Halley's Table, and this mixture of both circumstances led me to assume [the principle of] equally possible angles of inclination. Meanwhile it is unquestionable that the matter will clear itself up when this Table will be made more complete.[9] One and the same comet is not equally visible at each return, and one must enter them each time into the Table according to their best conditions of visibility.

The distance of perihelia from the sun may also do something with the distribution of the angles of inclination. A comet, like the one of 1680, must necessarily have a greater angle of inclination against the ecliptic because its ellipse is unusually narrow. It is generally clear from Halley's Table that the angles of inclination are greater, the closer the perihelion is to the sun. When the perihelion is farther, I find all angles of inclination in the Table. The plane of the orbit of a comet becomes all the more arbitrary, the farther its perihelion is from the sun. Its ellipse spreads out more, and all the easier can the comet keep away from the plane of the orbits of planets and in general from any other [orbit].

This is the solution which you, Sir, would propose to me as far as I can see it. Had I found it wholly correct, it would [251] have overthrown my rule, and I sacrifice this to the safety of our earth, which you, Sir, make much more free of disturbance, and also to the harmony of our ways of thinking, because I have more accurately demonstrated the rule which you have given, and I have removed the obstacles which made it rather improbable. It will always be a satisfaction if you show me new avenues as well as if you lead me back to them whenever I stray off in my ever bolder investigations. I can expect this because of your friendship, and each new proof of it will increase in me the most obliging gratitude into which you have so often already placed me, and with which I will seize all opportunities to show that I remain with the fullest loyalty, Sir, etc.

[252]

EIGHTEENTH LETTER

As long as you, Sir, keep giving such promptings for the continuation of my reflections on the arrangement on the world-edifice as the ones you have piled up in your letters up to now, I will not at all worry about lack of future insights, and it is a true joy that you always encourage me to explore even more the depths of the firmament. Material, contributions, questions, and doubts, in short, all that you wish to give me, I will try to utilize in such a manner that we may find even more, and that which has already been found may become more coherent. But you should not complain as if you yourself had not succeeded in finding something. The conclusions, which you draw from those which I have drawn by putting a dark body in the center of each system of fixed stars, and which at the same time you also make more general and extend to each Milky Way and greater systems, nay, finally, to the whole creation, are obvious proofs that you are not wanting in respect either of the universality or of the strength of mental powers and of their application in this subject. You reach out even to the most subtle questions and doubts, you seek means even to remove these in order to advance still farther, and you call upon me to do just that. You [253] request that I lead you, and lo, you bring me unnoticeably to the center of the whole realm of reality. Enthralled by the splendor, size, majesty, and beauty of that place, by the order and laws spread across the whole creation, to which each single part tends and around which they all direct their course and are held in well-defined orbits, you call upon a poet to lend his colors and vividly depict to us this place where the first driving wheel of each movement of the world-systems is located. From there you swing to the outermost borders of the world-edifice and reflect on the orbiting of the most distant systems around that place which ties together each single part through general forces and makes them into a whole.

But I return with you to the philosophical investigations. You are astonished at the enormous size of such bodies which keep in order entire systems of fixed stars, entire Milky Ways and, in brief, the whole creation. Although this astonishment soon leads you to the point of wishing to see some such bodies, I still think that you are not doing this for reasons which demand an autopsy in much easier matters for the sake of belief. You want to see because you think it possible that bodies of so enormous size, even though they are not more illumined than Jupiter and Saturn, still should have [254] such an apparent diameter that we could discover at least the nearest one with a telescope. You anticipate its phases and spots, and specify the changes in its apparent figure. It puzzles me that you did not exclude again the pale light in Orion from the class of Milky Way[s] and turn it into a brighter part of such a body.[1] You knew, of course, that Derham wanted to consider it not as true light but only as a kind of reflection which, as if through an opening, man sees as the splendor of the *coelum empyreum*.[2] The description fits more an illumined than an illuminating body. You knew just as well that one had found the apparent figure of that pale light changing,[3] which cannot very well be said in so few years about a Milky Way, or at least can only be explained if one could ascertain the reason for the greater or lesser transparency of our air [atmosphere]. The [particular] sun which illuminates such a [dark] body can appear to us as a very small fixed star, [and] if the body is great enough to have a notably great apparent diameter and is not too distant to have its light completely vanish through the ether, the telescopes will do us the appropriate service. This pale light is in addition in Orion, an area where you certainly wish to seek the center of our system of fixed stars.

[255] I have not yet seen that pale light through a telescope but only as engraved in a copper plate.[4] Since some fixed stars may stand in front of it, I cannot in spite of these reasons deny that this light should become too conspicuous to me, so enormous is the size of that dark body which I

put in the center of the system of fixed stars. Meanwhile, since it was found variable, it would take only a few years to see whether there appears in its variations something which could be explained from a motion around the axis and from the orbiting of a sun around it. I wished with you, Sir, that this be found so, or that such a body be simply discovered, because it would unquestionably be a considerable help for the closer determination of the system of fixed stars. I then would like to extend the proofs, which I have collected for the existence of such bodies, as you always wanted this to be done.

Since, however, this remains unrealized for the time being, I return to these proofs to demonstrate more exactly on what they are based and how far they reach, and here I once more find in the neat pictures which you, Sir, have drawn a welcome aid to direct my considerations in an orderly fashion. Now that it has been established that the fixed stars are subject to displacement, and since one can also consider it as equally well established that their motion is not rectilinear but circular [256] and that therefore central forces are present, the big question is the one which you, Sir, have already posed, namely, whether the regiment of a whole system of fixed stars can be entrusted to one single fixed star?[5] You have transformed this question into a similar one, when you removed the sun to consider how the disorder and scattering in our system of planets and comets should [even then] be reduced and eliminated. The question showed to you that Jupiter or, if it were removed, Saturn would ascend the throne because these two bodies have the mightiest forces at least among all those globes that come into our view.

I assume accordingly that a system of fixed stars is quite empty in its center, or that it has there no such body that could wholly rule and keep in order each star with its retinue, and consequently the following considerations impose themselves. Since the fixed stars should not move in straight lines, they must deviate from them and their velocity must be proportional to that deviation, so that they might always form a system. For this subordination cannot be eliminated; it is much too necessary for the arrangement of the world-system.

What can make the stars deviate from a straight line is nothing but the gravity [257] which they have toward one another, and the deviation takes place along the mean line of direction which arises when one puts together all the individual directions. It is therefore variable in each moment, and this already makes it very unbelievable and contrary to [the principle of] analogy. In fact, in our solar system this direction in the case of each planet and comet is rather simple with the exception of a few cases in which two comets pass very close to one another. Should in the case of fixed stars, each of which must remain much more undisturbed because of its large retinue, such a simple direction not occur, but another which is composed of millions of smaller ones and is variable in each moment? This hardly seems probable.

Furthermore it can very well be seen that in such a case the mean direction of gravity would, in the case of the outermost fixed stars of the system, be fairly regular toward the center, and that, because all the individual directions fall on the same side, the sum of all of them would be more considerable and therefore the orbital velocity, too, would be greater. But if I take a fixed star near the center of the system, it gravitates in all directions. The gravity in each opposite direction is eliminated, its effect vanishes, and the star has for all practical purposes no gravity or an unnoticeably small one at best. [258] Since then the star would not at all or only very slightly be deviated from the straight line, it could perhaps have no velocity at all. In brief, the order would not only be infinitely composite, but completely reversed. The outermost stars of the system would move the fastest, the inner ones would, however, move the more slowly, the closer they were to the center. All this changes and the order becomes harmonious and simple when I again put the dark body in the center.

To this I add that I am not inclined to leave in the same place a system of fixed stars. It belongs to a Milky Way and its center should follow a very orderly course around the center of that Milky Way. This is demanded by [the principle of] analogy which spreads from system to system and makes the laws of motion the simpler, the greater a system is. If you remove, for instance, Jupiter alone, how soon will its satellites be scattered?[6] If a whole system has to move in an orbit and remain in it, then there must be in the center a body which leads the system around itself.

About the size of such bodies I was not concerned at all. If because of the preservation of a simple and definite order they are necessary, they [259] must have their proportional size, astonishing as this size may appear to us. I find in general that in astronomy we must always get ready for greater things and leave behind the yardsticks which we use on earth.[7] These yardsticks shrink into the infinitely small and disappear from our eyes when we take an astronomical look at the firmament. Let us start with the satellites and go on to the planets and comets and from these to the suns, but the suns are in the idiom of heavens only bodies of the third rank.[8] With respect to systems we have so far fallen behind, but we must look upon this scale of ranking as something [that ought to be] more complete. Bodies have the fourth rank which rule entire systems of fixed stars, and these are the foundations of the Milky Ways of which each turns around a body of the fifth rank. And so forth, until we finally come to that body which rules the whole creation[9] as its own realm and makes it gravitate toward itself.

But I must show you, Sir, in yet another way that we still know very little what is great or small. As long as we do not know how far the suns are from the center of the whole world-edifice and in how much time they revolve around that center, we cannot say how fast the earth or any other planet moves. The ellipses in which we make the planets and comets move are mere [260] fictions and are no better than when we assume in spherical astronomy that the heavens revolve around the earth. Much remains to be done until the Copernican system becomes more than a useful and convenient hypothesis. You see, I now turn everything upside down, but I give you other and better proofs than do the Ptolemeans and the Tychonians.

To start with an example which is near to us and is already known, one assumes that the moon moves in an ellipse around the earth and goes around in that ellipse in about 27 days. This would be so if the earth remained at rest. But if one assumes that the earth moves in an ellipse around the sun, the ellipse of the moon disappears, and henceforth the moon moves in a cycloid [10] around the sun and somewhat faster than the earth because it has to take detours and still goes around the sun in the same time. This is again the case as long as one assumes that the sun is not moving. But the sun changes its position as does each fixed star. If now one assumes that the sun moves in an ellipse around the center of the system of fixed stars to which it belongs, the ellipse of the earth and the cycloid of the moon disappear and both turn into cycloids; [and] the earth moves in a cycloid of first degree, the moon in a cycloid of second degree. It is obvious that the velocity always increases [with the degree].

[261] This again will do as long as the center of the system of fixed stars, or the previously mentioned body of fourth rank, remains at rest. But since no body is at rest in the world, one has to assume again that this body has an orbit, and the sun therefore moves in a cycloid of the first degree, the earth in one of the second degree, and the moon in one of the third degree. But this remains always a hypothesis until we come to the body which directs the whole creation around itself. Were this of the 1000th rank, the earth would then be describing a cycloid of the 998th degree, and the velocity which goes with that would be its true velocity. How great is this and who will determine the nature [shape] of this cycloid, its thousandfold turns, and its amplitude?

What I said about the moon holds for each satellite; the statement about the earth applies to each planet and comet; and what I said of the sun holds true of every other sun, and so forth. The order will be the more composite, the lower the rank of a body, and the manner in which it is put together follows one and the same law, or the degrees of cycloids which they describe.

How does this please you, Sir? I readily grant you that I am bewildered when I want to picture to myself the consequences of this conception. But [262] you certainly know how to help yourself. Accordingly I will merely state them. In a proper sense our earth, and in general each celestial body, moves around the center of creation. The earth has nothing to do with the sun except to accompany it, remain in its neighborhood, and use its light and heat. For that purpose the earth uses a detour, because the law of gravity admits no other means to keep two or several bodies together. I will not determine how many bodies belong to each rank, but from each rank I take one to which our earth belongs and toward which it necessarily gravitates. It gravitates toward the sun to remain in a cycloid around it. Together with the sun it gravitates toward a body of the fourth rank, so that both may remain in its vicinity and in its retinue. This body, the sun, and the earth gravitate toward a body of the fifth rank which governs the Milky Way, and thus they do not disperse. In this way I march on, and in each step the cycloid of the earth takes on new and greater bends in addition to the smaller ones which it always keeps. It seems as if the satellites function for us as an example of what takes place on the large scale. The cycloids which we have so far assumed for them show us that we should not stay with the ellipses if we want to picture to ourselves the course of the celestial globes, and both [cycloids and ellipses] are much too simple [263] for bodies of so low a rank. The ellipses are applicable only to those bodies which immediately depend on the regent of the world-edifice, and are subordinated to it in such a way that they share among themselves its first commands and extend them farther through their immediate subordinates. These already have to make more effort and move in cycloids of the first degree as we have attributed this until now to the moon.

Sir, you see here a brief outline of the subordination which rules in the world. Was I wrong in saying that everything becomes simpler, the closer it comes to the whole? Or should I abolish my dark bodies which I set up as regents of the systems of fixed stars, of the Milky Ways, of systems of Milky Ways, etc, and introduce democracy[11] into the world-edifice? The latter is too big for that, and when not each part does without command what it has to do, more powerful forces must obtain to keep that world-edifice in a definite order and to make it through many degrees into a well-arranged and harmonious whole.

Since we shall in all probability never know the true [form of the] cycloid which the earth follows,[12] we have the choice in going as far with hypotheses as we wish. As long as we have to take into account only planets and comets, we shall depart [264] from the Copernican system for no [good] reason. It is much too convenient for these computations. Equally so can we apply it to cycloids of the first degree if we want to determine more exactly only our system of fixed stars. In a few thousand centuries the time will perhaps come to think of a cycloid of the second degree to bring into order systems which constitute the Milky Way.[13] So inexhaustible is astronomy that we must necessarily be satisfied with hypotheses and understand only how much these hypotheses must gradually be extended and made more complex. You go through un-counted degrees closer to the truth, and with each degree a new yardstick of space and time begins to be used. In the Copernican hypothesis our yardstick for space is the radius of the earth's orbit, for time a year or the duration of the earth's orbiting. Since we have hope to advance soon to the second hypothesis, we shall be able to take as yardsticks the sun's distance from the center of the system of fixed stars and the time of its orbiting around it. I do not specify the third epoch which requires even greater yardsticks, much less do I want to specify the epochs to follow.

Meanwhile, how convenient this choice of hypotheses is for us as we can always resolve the more complex into simpler ones. We already do this [265] with the moon for whose orbit we assume a cycloid of the first degree only when we are to explain the deviations and inequalities of its orbit caused by the influence of the sun. In each other case we let an ellipse hold good for it because the ellipse is simpler and we seek to graft on it the inequalities in an adroit manner to make the computation of its orbit all the more simple. In the same way we can let the orbits of planets and comets be ellipses, even though the orbiting of the sun permits us to discover there some inequalities.

We have already become completely accustomed to the language of the Copernican system, and it is now a much too convenient hypothesis that we should speak of cycloids where this is not demanded by special circumstances such as, for instance, the determination of inequalities in the orbit of the earth and of other planets, or the determination of the position of fixed stars which can become such a problem with time. In all remaining cases we stay with ellipses, we leave the sun at rest and use the so-far-customary language, because we can never use the true language in a more definitive way.

I will likewise resort to it now because I still have to answer the remaining points in your letter. You state, Sir, for good reasons that if the number of comets is greater [266], our sun remains more at rest and its center from the center of the system can deviate less than if the planets were alone. The sun has toward Jupiter and Saturn a gravity which is always very considerable and which would be sufficient to move it around in a circle whose radius would be about as great as the diameter of the sun. This circle would in addition be subjected to many smaller deviations which would approximately have a period of 20 years, like the time of a great conjunction.[14] But this does not take place because there is in all places around the sun a great number of comets, each of which pulls the sun somewhat toward itself, and because of this it does not for all practical purposes budge from its place. I always insist that the greatest comets do not come close to the sun and therefore within our view. We have derived this from the economy of space which must be used sparingly near the sun and from [the principle of] the undisturbed peace of planets and comets; it can also follow from the fact that the sun itself must remain undisturbed. In this connection you will remember, Sir, that I speak the language of the Copernican hypothesis. Considered in itself the sun is busy enough because in its cycloidal orbit of the 997th degree it must obey as many commanders. For I now stick with this subordination, however unsatisfactorily it can be demonstrated. It is too harmonious for me not to [267] assume it even 'without seeing' until [that time when] future observational evidence [will be on hand].

You have very much obliged me, Sir, with the solution of the question about the angle of inclination of cometary orbits. You have put, as is your custom, my investigation in greater light, and what I have found each time to appreciate in you so much is that you lead my investigation irrespective of your own opinion in such a way that you want to come to the truth even when you have to reverse your own thoughts. You can always rest assured that you are setting this noble aim without my suggestion, because I have in fact left the matter in doubt and the orderly agreement of your computations with Halley's Table was a chief objection to my conception. Assuming this to be correct, I would hold to the [principle of] various visibility and wonder whether it were possible to change an easy law into an easier one and to display in the Table the angle of inclination, whose number [frequencies] I let increase with its cosine, simply as uniformly distributed. Were they, however, to be in fact distributed in such a way, I would then fall back again on the perihelia because I plainly see that their distance from the sun has a very strong influence on the determination of the number and position of cometary orbits.

[268] Since numerous circumstances arise here and as, Sir, you yourself remark, the whole matter gives not the real but only the probable distribution of cometary orbits, I have not pursued any further the investigation of my questions, and I will let it stay there until the list of comets becomes more complete. It still may turn out that [if] the ecliptic has something special here it may be derived from the visibility or from some other cause. Perhaps the orbiting of the sun around the center of our system of fixed stars also has something to say about it. We can in general give very few reasons yet why the plane of planetary orbits, the plane of the sun's equator, and the plane of its atmosphere coincide.[15]

Because I mix this orbiting of the sun into everything that is connected with it, I have again considered a few days ago the zodiacal light or the atmosphere of the sun,[16] and I soon fell back again on the theory of vortices, which until now one eagerly tried to introduce into each system only to throw it away again and recast it in a thousand different forms.[17] That the sun retains its atmosphere, however great its speed of orbiting may be, did not puzzle me because we see this also in our habitat on earth. But that this atmosphere stretches from the sun [269] at least to the earth's orbit, and that therefore the inferior planets, or Venus and Mercury, move in it, almost led me to the conclusion that the sun carries along with itself, as if in a vortex, all material and bodies [that are] around it. I know well that one cannot consider the atmosphere of the sun as the vortex itself, and that a body which moves in it has its own proper orbit around the sun, which one cannot very well ascribe to the particles of the [sun's] atmosphere; for all that the whole[18] remains together and the whole solar system rolls on with the sun and its atmosphere. One has to view it as a very thin fluid material which coheres through its pressure against the sun, much as our atmosphere coheres and forms a body; otherwise I would not vouch against its dissipation, and it seems most natural to me that its form is not circular [spherical] but oblong [ellipsoidal]. The outermost point of the zodiacal light is often seen up to 100 and more degrees away from the sun. Whatever its shape may appear, it seems to me that one has to assume that the line drawn from the eye to that point is a tangent there. The atmosphere of the sun seems to run out into a point, because we see the outermost edge of it approximately as a burning glass raised on both sides when the eye lies in its plane. Draw now a triangle whose [270] three corners are the sun, the tip or the contact point at the zodiacal light, and the eye of the observer. This triangle will have at the eye an obtuse angle of 100 degrees, therefore each of the other two must be smaller than 90 degrees. However small you will make the angle at the sun, the angle at the contact point will always be smaller than 80 degrees. This is impossible if the figure or the outer circumference of the zodiacal light taken along its equator is circular and concentrical with the sun.[19] Therefore the side which is opposite to the obtuse angle is greater than each of the two others. Certainly, one correctly concludes from this that the tip of the zodiacal light must be farther from the sun than the earth, when that tip is farther away from the sun than 90 degrees. In other times, and presumably in other seasons, this distance hardly amounts to 50 or 60 degrees. In this case the tip is within the orbit of the earth. One has concluded from this that the atmosphere of the sun must be very variable, and this conclusion is necessary, [though] one can also assume that its figure is circular and concentric with the sun. But if one assumes this, there remains no means to explain how the outermost tip can stand 100 or more degrees away from the sun. The earth would then find itself inside this atmosphere because it stretches beyond the earth's orbit. From this, however, it necessarily follows that one has to see the zodiacal light stretched [271] across the whole sky, [but] this will not say more than that the zodiacal light completely fuses with the twilight. One will not see in it a well defined figure ending in a tip.

With us [in our latitude] this difference manifests itself according to seasons. We see the zodiacal light from the beginning of fall through winter until the beginning of spring. During summer it cannot be seen, and one largely blamed the twilight for this, but probably without a fully satisfactory reason. Because I take its figure to be oblong [ellipsoidal] and not concentric with the sun, I will consider it for the sake of brevity and greater clarity as an elongated ellipse in whose focus is the sun. The aphelion is outside and the perihelion is inside the earth's orbit. Our earth goes accordingly during the summer months through the sun's atmosphere at its aphelion. In the winter months the earth goes by its perihelion but outside it.

It will require several years of observation set up for this purpose to see how far the figure of the zodiacal light can be determined through such tangential lines, and how far the variability of the sun's atmosphere can have an influence on the apparent shape and length of it. It still seems to me most difficult to give a reason [272] why it is that the tip of this light can be 100 or more degrees away from the sun, if this is not to be derived from the elongated form of the sun's atmosphere.[20] Were, however, this figure elongated, I would not seek for it another reason than the orbiting of the sun and its gravity toward the center of the system of fixed stars. Still I leave the matter open for further experiences and observations. The theory of the sun's atmosphere does not seem to me to be sufficiently developed at least as far as this is known to me.

I have, Sir, brought up these difficulties from which I do not yet conclude anything definite, because I seek, as you can easily see, from every side new material from which to infer as time goes on something for the greater clarification of our position in the universe or at least to provide occasion for further investigations. If you, however, find about it and in general about what is presented in this letter a few lucky insights and more skillful presentation, I will be extremely obliged to you. I know that you never fail in that respect. I expect your comments with renewed assurance of the friendly devotion with which I am, Sir, etc.

NINETEENTH LETTER

Now we will be once more Copernican enough, or we will never be, or we should have never become one, for I am completely at a loss as to what to do after you, Sir, have talked to me of cycloids of the 1000th degree, of globes of the 1000th rank, of a thousand-part series of hypotheses, and of as many languages which one shall use in astronomy, and in all likelihood one must decide beforehand in each investigation and controversy the language one wants to use or in which tone to speak. How much the world is reversed![1] First, the earth was at rest. Then it started moving and the sun stood still. Now it, too, moves and the body of the fourth rank rests. But it, too, will be disturbed and rest will belong to the body of the fifth rank, and so forth until the order comes to the last [body] which is at rest not hypothetically but in fact.

It pleases me exceedingly that we keep the choice of putting any of these bodies at rest or in motion. This amounts to giving the pitch [to a band] for tuning up. If the earth is at rest, we stay with spherical astronomy to which all phenomena of the heavens [274] must be reduced. This happens through a translation from astronomical language into the common idiom. If the sun is at rest, we have to do this [translation in connection] with our planets, satellites, and comets, and we speak of ellipses and hyperbolas alone. Should one go on, it becomes the turn of our neighboring suns or fixed stars, and there come new ellipses and hyperbolas, and the language extends to cycloids of the first degree. Upon these there will follow cycloids of the second, third, and subsequent degrees.

But I return again to my first question whether we are now Copernican or what in fact is going on? If I myself picture the matter correctly, it will go like this. Ptolemy stayed with the popular or common language. Copernicus began to make the first step and taught us to recognize the alphabet of the first astronomical language. But he did not know that the language was merely hypothetical and would lead us only through many steps to the true language. Tycho Brahe misjudged the matter and replaced the vowel with a consonant, but one soon found out that the pronunciation suffered thereby and was not at all fluent.[2] Kepler and Newton have completely removed this confusion and found the means to raise the still rough language to its true refinement and to arrange their epic poems according to the mythologies of their times. But times change and our immediate descendants will permit this language only in poetry, [275] or will use for the abbreviation of expressions where it will have no significance. But where things matter, they will speak in a higher tone.

Thus I have now to develop the topic for myself. Insofar as Copernicus made the first step, there remains for us and for our descendants still a thousand other steps to take and even then we shall not be perfectly Copernican by a long shot.[3] Inasmuch, however, as we know that the last step will end at the body which directs the whole creation around itself, I think that we are thinking Copernican enough; or do you want, Sir, to go even farther? I cannot find at all where you would want to. Or should we have never become Copernican? But this could not be done otherwise than by firmly believing that the sun remains once and for all in its place and moves only around its axis. I grant you that this is an error which we must avoid because the sun remains as little in its place as do the fixed stars. The scene changes and there will come a time when the constellation of Orion will look perhaps as that of the Great Bear does now.

I prefer to remain with the Copernican idiom because it is the easiest to translate into the ordinary tongue and, as you, Sir, [276] remark, one will connect the next language to it in such a way that one reduces the anomalies of the cycloids to ellipses as we do this with the moon. I believe that astronomers on the moon can far more easily fall back on these various languages, although they have to take one more step because they live on a satellite. Their ordinary language sounds as if the moon had no motion at all. But since they constantly see the earth at the same spot of their sky, the earth has to appear to them as completely or almost completely motionless, or having no other motion than the one which it seems to have around its axis. It would be more natural to them to proceed to other languages and to establish the moon's motion around the earth. But they soon would have to let both the moon and the earth move around the sun, and on the basis of this twofold analogy they could doubt for good reasons the resting of the sun and would in the end stir everything from rest. Were there somewhere beyond Saturn some planets or comets whose satellites had satellites of their own around them, the astronomers on these latter bodies would have to make three steps before they could make the sun orbit, but these three steps would be more natural to them because their proofs would almost necessarily lead them farther out [into cosmic space].

Although I stay with the Copernican language, I still tried to form [277] for myself an idea of some of the subsequent languages to see something of the art of translating from them. I could compare with nothing better the various degrees of cycloids than with the waves of water. When set in motion by some cause its surface becomes uneven. There arises a series of waves following one another and between each two there remains a depression which will be filled again when the waves settle. This sequence of rise and fall forms a kind of sinuous line on the surface of the water when seen in a cross section and presents to me the image of a cycloid. If the waves are small, the rise and fall is also simple and here I have a cycloid of the first degree. If they become notably larger, a big wave shall consist of many smaller ones. The smaller curves remain but they will be pulled high up along the bendings of the greater wave and again back to the deep. It seems that nature finds great pleasure in such waves and oscillations, because we find them in almost all motions. Thus a ship moves on the sea through a series of such wave-like ups and downs.

The greater is the ship, the greater must be the series which can make it rise and fall. A smaller canoe follows the smaller and greater risings alike, and while the warship rises and falls only once, the canoe is subject to several smaller ups and downs which taken together [278] form the big wave. So does each celestial globe seem to move through the space of the world-edifice. All finally go around its center, but the greater one globe is, the less it yields and the simpler are its ups and downs. It is natural that all this should be more quiet and uniform there than on the sea. The universal wind which drives them on is more orderly and more consistent than the easterly wind between the tropics of our earth; and [its consistency] is more necessary for the preservation of celestial bodies than for that of a ship for whose replacement the new material is always ready to grow.

When I take as a sample the cycloid which our moon describes in the Copernican hypothesis, its bending is very small. The total length of a rising and a sinking amounts to about 30 degrees of our earth's orbit, and therefore to about one fourth of its diameter; whereas [the amplitude of] the rising and sinking taken together constitute hardly the 365th part of this diameter. The cycloid is therefore 90 times longer than wide.[4]

If I now assume the second language which allows the sun to move around in a circle, I change the ellipse of the earth's orbit into a cycloid of the first degree, and this must in all evidence be many times longer than wide. It still remains to be shown whether the sun has [279] traversed since Hipparchus' time, or in 2000 years in round figures, one degree of its orbit and how long such a degree is. During the same time the earth has already had 2000 ups and downs in its cycloid and Saturn 70 in its own. A comet that returns only in a few centuries has an even smaller number [of such ups and downs]. Regardless of this I would like to draw up all cycloids according to length. All depends on the velocity of our sun. If I set it at only a few times greater than the velocity of the earth in the Copernican system, the cycloids will be very elongated and their bendings [amplitudes] very small.

It does not, however, appear that this should necessarily be so, and the satellites of Jupiter and Saturn describe in the Copernican system such cycloids which do not lend themselves to a comparison with the waves of water. Because they move faster in their ellipses than the planets move in their own, they move in fact backward when they come between the planet and the sun, and their cycloids cross one another.

I cannot present their figure in any familiar image and I am really getting unwittingly tangled up in all these orbits. The lines which the calligraphers draw around their letters hardly seem as complicated as do the true orbits of the celestial bodies. I wanted to follow up [280] the course of one of the comets whose orbit around our sun at rest is a hyperbola, but I did not succeed and had to satisfy myself with the notion that it is a body which the sun somehow met in its orbit, bent its path around itself, and let go to seek new fortune. I almost forgot the splendid vision which we entertained previously.

How wonderful should things look in the world-edifice! I find there two classes of order. In the course of things on our earth the disorder is apparent because on a cursory look everything goes confusedly. If one considers the whole, the general laws, and with these order, come to the fore. In the firmament the reverse is true. The simple order is the apparent one. In fact what can appear more orderly than the daily course, the rise and setting of the sun and of each fixed star? Once a day the whole heavens turn around the earth, and at night the moon seems to be the leader of the whole host of stars. If one takes enough time and leisure to look into this scene carefully, small deviations begin to show. The moon wanders backward from star to star, and the planets, too, begin to show that they follow their own mind, strongly as they may be dragged on by the general stream. To remove this disorder the observer [281] begins to think Copernican and brings planets and comets into such an order that could not be of greater excellence. But gradually the fixed stars, too, show that they are not dead masses but have their own life and motion. This is a new reason to view also the [Copernican] order of planets as fictitious and to conclude once and for all that the true order is the most complicated one and will never be reached by us. The world-edifice, considered as a whole, has a thousand driving wheels mutually so oriented that each bigger one locks into the nearest smaller one and this into the next one and causes in the smallest one uncounted wobblings.

Time and space are so interconnected in the world-edifice through the motion of each body that the apparent order should be the simplest and that, as we must take a more complex order, the time whose duration we want to picture also becomes greater. In what an orderly and precise fashion does the daily rotation measure for us the days and the hours and its parts which we have to use in every business! No clockwork is more exact than the daily course of each fixed star![5] But there is a need for periods of longer duration. If the course of comets and planets which form the earliest and most obvious exception in the first order should be measured, we must then take into account the next driving wheel and we can stay again [282] with that supreme order which Copernicus traced out for us. If, however, periods of many centuries present themselves, new deviations will be in evidence which one has to bring eventually into a new order. One has to proceed to the third driving wheel which measures its rotation in Platonic Years.[6] Thus we have through apparent orders precise clockworks for hours, days, years, millenia, and for all further periods which step by step become a million times greater than the immediately preceding ones.

I am once more astonished at this arrangement. The most composite of all orders must show us a very simple form by which we can measure, divide, and apply our time correctly. How necessary for us is this disposition of the Creator! Farmers had to teach us astronomy because it was necessary for them to know the season for each work on the fields, not to sow in an inappropriate time, [and thus] to avoid general famines which in ancient times occurred so often simply because the expectation of a ripe harvest failed due to lack of knowledge about the seasons. Perhaps there will be reasons to know the millenia as accurately as we now know the seasons, but I am not concerned about these periods, and the matter can be left to curiosity. It increases in me from day to day and [283] the more I consider the world-edifice, the more I am faced with questions which I would gladly explore.

Meanwhile, it is enough for me to understand the arrangement of the whole, undetermined as particular details may remain. Where are we now in the world? How far is our earth from each of those bodies around which it swings? How many individual turns does it use with respect to the next greater one, and how many of these with respect to those still greater? How close can the earth come to the center of creation and how far can it recede along its cycloid of a thousand greater turns from that center? At what point of each turn does it move at this moment? How great is its true velocity, or is it at this moment completely at rest as it begins to retrace its path, or would it have its greatest possible velocity, which is formed by the sum of velocities in each ellipse which the bodies of each rank to which it belongs would traverse, if the body of the next highest rank remained at rest? Does not this velocity become almost infinite, or will it be controlled by the fact that some of these bodies move forward, some backward, [and] still some others upward and others downward? No enthusiast of fireworks can place more wonderful curves into the air than the earth or any other celestial globe in the depths of the firmament. And in that unthinkable curving of the earth's orbit there must, [284] nonetheless, appear before our eyes the simplest and, for the first use of reason, an equally beautiful order.

You may see, Sir, from this how I am getting used to making a whole list of questions. But I was no longer frightened. However extraordinarily the celestial globes may move across each other's path, I have no worry at all about their being derailed. I perfectly understand how there was necessarily given to each of them at the moment of creation a mass of definite size, a degree of velocity perfectly proportional to it and to all bodies to which it is subordinated, and a proper direction, so that each planet could have through that infinitely complex order a perfectly correct clockwork which showed its time spans of each duration, and as the time spans increased, new driving wheels unfolded for the purpose of measurement. For I am even more astonished that in all this only the simplest order is apparent, and the true order is as complicated as it can ever be. We can wholly overlook the true one, and it would not serve us even if it were the simplest of them all. But what is so convenient for our use and is needed by us so much presents itself to us under the simplest form. Perhaps this form is so essential that merely to maintain it the true order had to be [285] very involved. Even the suns around which planets and comets must move had to be heavy or, to speak for brevity's sake with Newton, have an attractive force. This force immediately takes away their rest, little as it could assert itself in the beginning. To bring the motion into a simpler order and to make the whole enduring the suns had to be grouped in classes, or distributed in systems, and a regent had to be given to each system. But since there were several systems, the difficulty arose again, and the regents had to have an overlord who directs them with their retinue around. All this was still too little, and this subordination had to have several degrees. The more degrees came to be added, the more complicated became the path of the prior celestial globes; only the appearance remained simple, as it had to be if the firmament had to present to each planet the most perfect clockwork.

You have now led me, Sir, into the capital city and you let me see the outlay of the whole. But do you know what is still missing for me? Do you remember that you reproached me as if I did not want to believe without seeing? In fact, what do you hope to make of the pale light in Orion? You pile up so many considerations upon one another that I will soon see it as the clearer portion of the regent of our system of fixed stars. For I firmly believe that you are serious about it [286] and you now merely look for other reasons to make the autopsy appear necessary. Did not Derham want to see this not as a proper light, but a reflection?[7] It appeared to him too uniformly bright and so carefully cut around as if it were an opening through which one could look into an illuminated place. What can fit more accurately an illuminated surface whose reflected light is still weakened through the ether and because of this it becomes even more pale and more uniform? Furthermore, has one found its shape changed? Should one derive it from the opacity of our atmosphere that the edge of this light does not gradually disappear into the dark but is distinctly separated from it? I really regret that it will take more observations to conclude something from these changes. I would gladly seek some way out of this problem.

Still, I will persevere. But the reason, which you, Sir, derive from the apparent size of this light to see it not as a light from your dark bodies, you have perhaps adduced only because I was preoccupied with this size in my previous letter. I have only drawn from it the conclusion that one should be able to spot at least the nearest of these bodies with a telescope. Consider now only what a mass belongs to a body which has to keep in order a whole system of fixed stars! [287] You tell me in that connection that we still do not know what is great or small, and that we must familiarize ourselves with ever greater things. What is the earth compared with the sun, and what can the sun be against such a body? The diameter of the sun is well over twice as great as the diameter of the moon's orbit around the earth. Could not the cross section of such a body still be greater than the orbit of Saturn around the sun?[8] I should think then that it would be visible. In all likelihood we are hardly as far from it as the distance of some of the fixed stars, and the fixed stars which one sees in that light in Orion must necessarily be behind one another at unequal distances, because they appear to be too close together. But I will leave this with you to observations, and I hope that time and circumstances will permit us to set them up together.

Meanwhile, I cannot yet abandon the possibility of seeing such a body, and if the light which it reflects were not in itself so weak, [and again] if this light were not weakened too much in that immensely long road which it has to traverse through the ether, and finally if our atmosphere did not possess at night a certain degree of brightness due to the light of fixed stars, then I would be bold enough to believe that we could see also the body that rules our own Milky Way and in general all those bodies around which the earth bends [288] its cycloid, and that their apparent diameter would still be big enough for that.

My proof is taken from analogy and with its help I conclude that the body which rules a system must have a diameter which is still very visible through a telescope even from the outermost confines of that system. In the case of the satellites of our solar system this is obvious. They see their planet so big that it can illuminate their nights. Saturn, the outermost of planets,[9] sees the sun under an angle of 3 minutes. The comet of 1759 in its greatest distance sees it under an angle of one minute. A comet which is 60 times farther away could still see the sun under an angle of one second,[10] which can be magnified with a good telescope to 2 and even more minutes. I scarcely believe that a comet, which is to return or moves in an ellipse, would recede so far. The shortest time of its orbiting would be over 35,000 years, because it would stay out 2200 times farther than the earth.[11]

Again, the attracting force of a body decreases as the square of the sine of its apparent radius.[12] If it is not completely unnoticeable or so minute that it no longer counts, then one cannot assume this [289] apparent radius to be unnoticeable, [and therefore] it still must have a recognizable size. For you, Sir, admit that a body which should rule such a system must impress with its power to the limits of its system, or conversely that its system should not extend beyond the more powerful part of its sphere of influence. All the more should it have a considerable diameter still visible at the periphery of the system.

You can easily see where all this leads. Our earth belongs to gradually larger systems. The earth is therefore in the sphere of influence of each body that rules it; thus the earth is in the sphere of influence of the sun, in that of the regent of our system of fixed stars, in that of the regent of our Milky Way, and so forth. Each of these bodies must therefore occupy a still recognizable part of our sky and therefore be visible, at least through telescopes if no other hindrances are present. But I do not think that we can see more than one of them, however good service telescopes may render, because the weakening of light through the ether and the nocturnal brightness of our atmosphere remove from our sight a light which is only reflected and must cover such an enormous route.

What other remarkable things do you still give us to discover in the firmament? One often finds only [290] when one approximately knows what one should look for. I am more and more inclined to believe that Derham would not have thought even once in his dream of the *coelum empyreum*[13] had he known that one had to look for bodies of such size even behind some fixed stars, and just as little would I have thought of a Milky Way [there]. But now the matter is firmly settled with me. I will seek time and opportunity to make something of this discovery, and should I unveil the regent of our system of fixed stars I will grant you everything else, unusual as its size may appear to us.

So much I understand rather well that through the manner in which you, Sir, present the world-edifice everything becomes simpler, the closer one comes to the whole. The ellipses, which we admire in the Copernican system in connection with the planets and comets as an unusually orderly and simple arrangement, do not vanish thereby from the world-edifice. Being the simplest they will only be placed to worthier places, and the bodies which directly depend on the regent of creation move now along them with simple and solemn majesty commensurate with their rank. Next to these follow cycloids of the first, second, third, and following degrees, until we come to the cycloids of satellites. So extended is this simple law over the whole through each degree that we hypothetically assume such an order,—which rules in fact [291] from the center outward and extends itself from the first regent to its immediate subordinates, from these to the following, and so on,—in the case of each subordinate regent, even if the latter were the thousandth in the ranks, and we let its immediate subordinates wander along the simple and comfortable ellipses.

I readily grant you, Sir, that this order is so enlightening to me that I cannot conceive of anything more harmonious and more complete. It resembles a series where each member is formed from the immediately preceding one through a single law, and among those series which are called *recurrentes*[14] this is the simplest and most perfect. I cannot pause here long enough, and I would regret it very much if the regents were to be rejected only because of their enormous size.

I take, for instance, the series 1, 1, 2, 3, 5, 8, 13, 21, 34, 55 etc., subtract each member from the next in order, [and] there remains 1, 1, 2, 3, 5, 8, 13, 21, 34, etc., which again is the same series and thus permits again the same transformation. But each time the first member falls away. It is much the same when I put at rest a body which I assumed to move in an ellipse. Therefore each of the cycloids depending on it becomes simpler by one degree, but the series remains. We can carry this on as long as a body of a [292] given rank remains at rest or moves in an ellipse. We do exactly this in the Copernican system where we let the sun remain at rest, so that the planets and comets may move in ellipses.

What you, Sir, say about the shape of the sun's atmosphere seems to me worthy of investigation. The various descriptions which I read about it give me yet no satisfactory account of why its outermost reach, or the tip of the zodiacal light, can be at a distance of 100 degrees from the sun and appear clearly outlined, if its true shape were circular and concentric with the sun. This shape should lend itself to a more exact determination through the tangents which you draw and, if memory serves me right, it has already been noted that the apparent shape of the zodiacal light closely follows our seasons.[15] I, however, believe that we in Europe never see it in every part of its [whole] extent. In fall one sees it only in the morning, in spring only in the evening, therefore in both cases [one can see] only the part which is opposite to the colure of the summer solstice. In winter it can be seen in the morning and in the evening, and consequently that part of it [can be seen] which reaches out toward both equinoctial points of the zodiac. The fourth part which is opposite to the Ram does not come into our view, and in summer when [293] the earth is in that sign we see nothing of that atmosphere. Its inclination against the ecliptic is set at $7\frac{1}{2}$ degrees and its nodal line goes through the 8th degree of the Twins and Scorpio. These are exactly the circumstances which cause the earth to be found in the summer always within this atmosphere, so that its light spreads across the sky and, if I am not mistaken, some French astronomers in China preferred perhaps for this reason to present this light as a second twilight.[16]

If this atmosphere were more extended toward the Ram and therefore also toward Orion than on the other side, then I would find another new reason for looking for my prospective dark body in Orion. This atmosphere would have something similar to the tails of comets which almost always point in a straight line away from the focus of their ellipses. The observations which are still required might, as in the case of the light in Orion, be carried to completion in a few winters. I am so eager about both that I would do everything to establish them soon, and I hope that I will be able to do this with you, Sir. We shall now look at Orion with eyes other [294] than before and with an incomparably greater satisfaction once we have found something there.

Do you, Sir, seriously want to bring up the vortices again? Should a vortex have in its center no great body like a regent? If this is not necessary, I again lose all my hope to discover something in Orion. But you will not leave a vortex to itself. I believe it would be disorderly, like the vortices which form in air or water and dissolve again. The order in the world-edifice must be more definite and uniform. It seems to me that the invention of the vortex is a precarious thing, and I know that you, Sir, would rather stay with the bare law of gravity as a correct experience rather than keep concentrated on a mechanism which one could not guarantee, because one will always doubt whether several others are not still possible.

But I must conclude in haste. Sir, you see how I tried to present your system from every angle. Perhaps I gave free rein to imagination, for astronomy excites me more and more. But I know that [295] you will always lead me back whenever I go astray. I am looking forward to your thoughts with expectation. Nothing would please me more than the opportunity to proceed soon to the observations which I am now planning. In the pleasant hope that they will be carried out with you, Sir, I remain with sacred devotion, Sir, etc.

[296]

TWENTIETH LETTER

You have, Sir, every reason to ask me how much farther I still want to go if it occurred to me that all this was not enough. The world-edifice is now an interconnected whole, arranged harmoniously according to a general law. Do you want that I should still patch to it pieces that do not belong to it? And truly, have I not already rambled somewhat beyond the limits of what is credible? I drew conclusions, and surely enough without having in each case appropriate observational evidence on hand and without knowing how far they would lead. But your last letter which pleased me very much and obliged me in every respect presents to me the whole in its interconnection and from all angles, and shows me how far I have gone. Is it not, in fact, unbelievably far?

About the clear presentations, which you, Sir, offer about the arrangement of the world-edifice, it is certainly not true that they are not probable. You show the order which extends in it through a thousand degrees with such vivid and captivating strokes that one is almost tempted to ask whether another can be true without again rejecting the [principle of] analogy and, if [297] all this still remains in doubt, one can still say with the Romans: *se non è vero, è ben truovato* [even if not true, certainly well said].

In the study of nature it is difficult to draw correct conclusions from empirical evidence, especially if one is to erect a system without hypotheses. How much more difficult ought it to be when one does not consult the empirical evidence but satisfies oneself with general considerations! Such is the case in which I find myself. I am drawing not merely one conclusion but a long and multiple series of them. Such a series resembles extensive calculations in which it is advisable to set up a probe from section to section [to see] whether one did not miscalculate before one proceeds with his calculations into the unknown. Empirical evidence should be such a probe in respect to my conclusions. But I satisfied myself with only indicating them and went on to conclude as if everything was beyond doubt and as if each empirical evidence had long since been provided.

I can therefore look at my conclusions as an example of no small boldness all the more so as we live in times in which the freedom to arrange nature according to one's own conceptions is wholly banned.[1] And I arrange not only individual parts but the whole of nature, [nay] the whole extent of creation according to my conception. Can one be any bolder? Who gave me the balance [298] to test each reason and to weigh exactly each conclusion, or to place on it the celestial bodies to see how much weight I should give to them, since I have not hesitated to turn the earth into a speck of sand? Can the reason that we do not know what is great or small be the justification for putting bodies of any size there where one needs them to carry the system to completion? Are those bodies just as necessary and is the system firmly proven to the point that one cannot leave them aside? Should the edifice of the world have such cornerstones and pillars to make it enduring through all ages? Should one grant [the existence of] those bodies without seeing? Who was beyond the Milky Way to take a look at them and bring home the news about their size? Whence the right to present speculations as binding conclusions, and fantasies as truths, and in particular [whence] the right to put big bodies there where we can never see them and where, consequently, we shall never see whether or not the very opposite should be the case or something entirely different? Where are the proofs if we believe without seeing and if, in its place, they should perform a similar service?

You see, Sir, I now step before the judgment seat of reason. This is her voice and she demands the proof of my system and of my reasons as her own property. What is erroneous she will reject, what is exaggerated

[299] she will properly retrench, what is immature she will refer to more empirical evidence; she will approve what has been demonstrated, she will give strong support to what is demonstrable, she will fill the gaps, connect each part, extend the limited to the whole, and bring it into a universal and firm interconnection.

So equitable is this judge where she rules that I step before her with pleasure, for I reject all arbitrariness, wholly submit my system to her enquiry, and where I erred I can expect truth in place of error. Each time this exchange is welcome to me and becomes always more so. Reason hates the false decor by which advocates of error try to make her [look] more impressive. She recognizes that one should doubt where reasons are not sufficient, that one should present each reason stripped of all that makes it apparent, and then she will assist if proper tools are on hand, and reserve for herself to pronounce the sentence, or if she suspends it, to indicate what is required to proceed to the conclusion.

You have, Sir, given me so many promptings for the construction of my system that I did not know in any previous letter what ideas would occur to me concerning the next. And in this way I came unnoticeably further, because [300] you yourself indicated to me how far my conclusions would lead me. Now I look over the whole and find each step that I have made. I will now present them in their strength and weakness. Judge it then, Sir, how far they reach. You have submitted to reason the complete arrangement of the system of your thoughts. Your judgment cannot differ from it, and I will see how far my world-edifice will be approved or changed after each requisite inquiry. See now the list of questions into which I clothe that system to be scrutinized before such an esteemed tribunal.

1. Are the fixed stars moved by central forces?

    That they move is [clear] from experience. This is also the question whether their motion is rectilinear, or curved and orderly.

2. Does the Newtonian law of gravitation extend across the whole world to make it an interconnected whole?

    The question is whether gravity is a property of matter or belongs to the [pressure of the] universal ether. Moreover, the world ought to be a whole [301] and not a patchwork, and its parts should have a closer connection which includes time and space.

3. Should one divide the Milky Way into individual systems of fixed stars, or do the fixed stars outside the Milky Way constitute such a system?

    The appearance of the Milky Way gives the reason for assuming the former, because it appears in several places to be divided into small clouds. The latter seems to follow from the fact that the edge of the Milky Way is too clearly cut.

4. Has the sun an orbit of its own?

As well as other fixed stars.

5. Are there small deviations in the annual orbiting of the earth and of the planets, and does the displacement of nodal lines and aphelia arise because of this?

This occurs with the moon in reference to the sun. It should also be so with the planets. One must, however, see whether the deviations are observable and what should be ascribed [302] to the comets in connection with nodal lines and aphelia. The steady part there might be derived from the orbiting of the sun.

6. Are the true orbits of planets and comets ellipses?

They would be if the sun remained at rest. In fact, however, they are cycloids.

7. Could one still retain the ellipses?

Just as well as in the case of the moon and of other satellites.

8. Is there in the center of a system of fixed stars a body which keeps it in order just as the sun does with its planets and comets?

This seems to be so according to [the principle of] analogy and [thus] the order in the orbiting of fixed stars of the system becomes simpler. It remains to be seen whether this is not the only means of maintaining a stable condition.

9. Should such a body be great and bright?

[303] It must have a considerable mass. The size depends on the density of matter. Both should be proportional to the size of the system. The body can have a weak light of its own, or it will be illuminated from a sun near it. If only the suns are destined to give light, the body in question remains dark and will be illuminated. [2]

10. Can one in the long run discover the body which is in the center of our system of fixed stars?

If its mass is not many times denser than that of the sun or of the planets, this can happen because it will have a considerable apparent diameter even at the borders of the system.

11. Would it have phases?

[It would] if a sun orbits around it and illuminates it. And the shape changes if it has spots and turns around its axis, as this [rotation] appears to be universal.

12. Does the weak light in Orion show such changes and can one consider it to be such a body?

[304] Experience must teach this. If the changes follow simple

or double periods and according to the theory of phases, it cannot be explained otherwise.

13. If the systems of fixed stars have such bodies for regents, would they not constitute together a greater system in whose center there is again a regent which extends its sphere of influence through this greater system?

> The systems taken together form the Milky Way. Should it happen that each or at least one of them has a regent, then the analogy can certainly go on and the Milky Way also has a regent, which rules it completely and directs it around itself.

14. Does not this regent of the Milky Way have a still more considerable size?

> Its size must be proportional to its domain.

15. Is this regent the last from which one cannot go further?

> [305] The Milky Way still can belong to several other Milky Ways and to an even greater system. If this is so, the analogy extends further. It, however, begins to become arbitrary because it moves beyond our sphere of vision. The analogy will to all appearance have more degrees. But the first degrees must first be demonstrated as much as possible from experience together with what concerns the regents of the systems.

In such a way I thought to capitulate if the investigation of what is certain and uncertain in my system were to be undertaken more rigorously. I could have piled up still more questions, but these seem to me the principal ones and most conducive to an inquiry. You, Sir, are most willing to dwell on the subordination, which goes in this system from step to step and leaves for us the convenient choice of making ellipses out of cycloids of each degree, and which unveils for each period of time and its measure new driving wheels of this most perfect clockwork and lets us use it. You have shed light on both conveniences in such a vivid manner that I can add at most only shadings to it if I wanted to pause longer on the subject. I had to show that any other arrangement would be neither so perfect, nor so simple, nor so convenient and [306] harmonious as this and that the analogy would completely vanish if the larger systems were not arranged according to the pattern which we see in the smaller ones. This seems at least to be necessary, because in all of them some central forces are at work, and because the order in the larger ones should be simple rather than complex. As long as one considered the sun and each fixed star to be at rest nobody was upset by their size. Their resting seemed to be commensurate with their majestic size. Were even greater bodies to emerge on the horizon, they would have even greater claim to that rest. I cannot at all part with the resolve to make the motion the more uniform, the greater the body and its retinue.

The computation of the manner in which the fixed stars of a system might revolve without a regent around a common center would mean endless complexities, if one were to take each case into account, so great is their number. If one takes only two bodies that move in the same plane, there is certainly the possibility to direct them in ellipses or circles around the common center of their gravity, so that they always keep the same orbit around that point. This center will be the focus of both ellipses. Their major axes lie in a straight line, but the aphelia are opposite [307] to one another and the [two] bodies should move in such a way that they are always on a line passing through the common focus. Thus they come around in equal time. Even when three or several bodies are to move in concentric circles, they can be so arranged that because of the central forces they move around in equal time. But such a perfect equality of time is well-nigh impossible in nature. If, however, one takes the periods to be unequal, or should the motion take place in different planes and directions, then I would not be able to assure that order and stability might take place there. The problem is always given better care when the bodies have a body in the common focus, as the sun is with respect to the planets and comets, which keeps them all, numerous as they may be, in a definite order. And I can ask, Sir, with you: how would a comet without the sun be permitted to approach Jupiter?[3] I have no intention of explaining the motion of celestial bodies by vortices even if I admitted their existence, because we had so little success so far in fathoming their mechanical arrangement. Here, too, as in the study of nature in general, the matter hinges on analytical conclusions, which one has to use if one without assuming hypotheses wants to bring out the true arrangement from what empirical evidence teaches us. Outside astronomy we have yet very few examples of this method, and one is yet hardly familiar with it, [308] so that when Newton used that method to discover the law of gravity most others viewed it as nothing better than an arbitrary hypothesis, and those who felt the strength of the proof always had to remind themselves that it is not a hypothesis but an empirical evidence. Newton himself does not seem to have perceived the universal aspect of his method, well as he knew that he could build on his conclusions. And because he labeled it with the name attraction, he himself provided reason that people counted his discovery among the wholly outmoded occult qualities and were all the more disposed to reject it. Since, in addition, he left the space empty, the Cartesians who tightly filled everything had even more apparent reasons to look upon the whole discovery as something arbitrary. They had their vortices too close to their hearts to sacrifice them too easily.

It is true that one does not get too far with this method and one gets very impatient with it, since with hypotheses one can extend conclusions throughout a whole system of thought, and in the course of one's satisfaction one can grow unmindful of the fact that in a few years the whole system can go the way of all fictions which one at most reintroduces only to reject them again. In this respect Aesop's fables have a considerable advantage over physical hypotheses. The former contain [309] the ingenious in fiction and in moral lessons, the latter simply collapse as soon as one realizes that they are not true.

I believe, meanwhile, that a mechanism certainly is at play in the motion of celestial globes, and it ought to be firmly demonstrated that bodies can in a completely empty space and without intervening matter influence one another in such a way as Newton's followers take [gravitational] attraction.[4] We are much too accustomed to presupposing in each effect an indirect or direct contact to dispense with it on occasion without strong proofs.

In whatever way a vortex is assumed, I doubt very much whether one can leave its center empty. One needs vortices to account for two experiences. Either the vortex, for instance that of the sun, serves only the purpose of pressing the planets toward the sun, and in this case the planets have their centrifugal force or their velocity on their own and regardless of the vortex. Or the vortex has to lead the planets completely around in a circle, and in this case the vortex duplicates the circle because its material already takes that route and like a stream carries the planets along. This would do if the motion of each planet and comet were circular and concentric, [310] and all would lie in addition in the same plane, and all would be driven around in the same direction. But since there are as many comets moving backward as forward, the current of the vortex should consist of as many streams as there are orbits of comets and planets, and since you, Sir, have juggled their elliptical spokes across one another, these streams must flow through one another and the velocity should conform at each point to Kepler's laws. What is going to give to these streams course, order, and direction? This question comes back again and again.

If, however, one changes the current into pressure, this would either have at each point the direction which the current should have to drive around the body if this had no velocity of its own. Here again there is nothing simple and the former question comes back again. Or, if the pressure is simply toward the center or toward the sun, we would have no vortex in the proper sense. The pressure could manifest itself even when matter remains at rest. The question would now be about the various degrees of its density and resilience which should increase with distance from the center. I know well that enthusiasts of the vortex would like to explain this pressure through various kinds of motion. But I do not believe that we have yet enough empirical evidence [311] to establish here anything firmly, and the explanation of how this pressure should materialize will always be the rub of the difficulty whenever that pressure is to be strictly demonstrated.

Meanwhile, let that pressure be taken as a foundation, and let gravity be ascribed to a material [medium] that deviates the planets, which have their proper velocity, from the straight line toward the sun. Then one will come up against the question whether this material which manifests its pressure in the sphere of influence of the sun goes around with the sun in its orbit, or whether the sun, wherever it may be, does in fact assert itself like a sphere of influence. In the latter case the pressure of the medium or of the vortex depends on the sun, and consequently I put into each vortex a body which rules it and influences it in all points. In the former case we fall back again in all appearance on currents, and we must assume them also with the planets and comets, with the difference that in their case only a part of the current forms the vortex. The other part is in motion because it must turn [in that vortex]. The question whence the body has its motion will now change into the question whence the vortex has it, and is just as unsolvable as the former question. Since currents are so clumsy, I will stay with the former case and make the pressure of gravity dependent on each body, [312] however little we can explain the mechanism implied. [5]

The pressure is so exactly proportional to the mass of the body that I would not admit a vortex without a body. The vortex of a whole system of fixed stars would have to be of a very different kind if its pressure did not depend on a body toward which each fixed star would gravitate, as planets do toward the sun. One has invented the vortex to spread across the whole universe a uniform arrangement. But the time does not seem to have arrived yet to think about its mechanism. Several kinds are possible, and we have but a few data to make a choice.

Since our sun has its own course and presumably along a very complicated cycloid, the Copernican world-edifice cannot very well be demonstrated otherwise than as a hypothesis, and our spherical and theoretical astronomy differ only in that in the former one puts the earth at rest, in the latter the sun is at rest. I wonder what an astronomy would look like which has been rigorously demonstrated in every respect. It seems as if one can get to the truth only through a series of hypotheses, and that one has to reject each previous one to espouse the next and to abandon it again. This is at least the road along which one has [313] indeed traveled, and the question is whether another road can be found. Each of these hypotheses justifies itself by presenting the appearance of the case from a series of viewpoints, and it is unquestionable that one can certainly assume a hypothesis, but one cannot extend it further than permitted by the appearances on which it rests. One assumes that the whole heavens rotate around the earth, and from this one has to derive that the earth rotates on its axis. This is certainly true. But since this is not sufficient, one assumes that the earth moves around the resting sun, and from this it must follow that the sun is not at rest but changes its place and so forth. It is not clear for me whether one can handle all this without mechanical or cosmological reasons, or whether one should help oneself by piling up fragments of individual proofs which, though they put the case beyond doubt, in respect to coherence and rigor should be much more beautiful.

Here one can help oneself quickly with cosmological reasons. Assume, Sir, that there is nothing in the whole world-edifice at absolute rest, and we shall stir the earth, sun, and fixed stars alike, and they must move around. Assume further that in these motions there should be present all possible diversity with respect to time and space, and immediately the earth rotates on its axis and forthwith the rotation of the fixed stars [314] in 24 hours vanishes, and the distance of each celestial body becomes different. For when you assume that there should be present order and general laws, there arise systems before long, the planets are brought into one class, and the fixed stars themselves become distributed into clusters.

I will not guarantee whether one will consider these conclusions true if they are not also proved from other reasons, as there is very little respect for cosmological propositions. The mechanical proofs have already been explored. The flattened shape of the earth is an incontrovertible evidence of its rotation on its axis, and I do not see what can be criticized in that. Here belong also the elliptical orbits of planets and comets around the sun which, if one puts the sun at rest, can be demonstrated mechanically from the law of gravity. But now I cannot say according to what order this proof should be presented. The orbiting of the sun is certain enough but not yet demonstrated a posteriori, which seems nevertheless to be necessary. The ellipses are true only hypothetically, and the question is how one can prove them without some presuppositions.

It seems that in astronomy we have come up luckily to the right path, and we have stretched the view as soon as we saw it in its first form [315] so far out until it began to reveal itself under ever new forms so as to be stretched even farther out. And we probably must abide by this condition because still further expansions of that view will be required.

Meanwhile, Sir, I cannot with you be amazed enough by the fact that this view shows each time a simple order and is so convenient with respect to the astronomical usage we make of it. The difference, which we find between this view and the course of events on our earth where the view reveals mere disorder, arises mainly because the firmament is arranged according to a law which makes the motion more complicated with each degree but always in a similar manner, as you explained by your recurrent series. With us, however, there are more and very different laws, causes, and mainsprings, which are so intermingled with one another that it takes much time and the survey of a whole series of changes until something universal unfolds itself. In this respect we know the heavens better than the earth, and perhaps we shall sooner bring the fixed stars in order and be [sooner] able to determine the changes of their paths than the changes of weather and the movement of the barometer. Here, there are too many causes, and each depends on uncounted circumstances [316] and, one may even say, on every irregularity of the earth's surface. Each mountain, each valley, each peculiarity of the soil contribute to this.

From this we can conclude that the firmament must be more enduring than things on our earth and the empires of the world. The firmament must be a clockwork which has new driving wheels and pointers for each time span, and this clockwork must serve each celestial body in an appropriate way. Each driving wheel must emerge only when the already elapsed time begins to have a noticeable proportion to its period of revolution. We know at most the seconds and minutes of this clockwork, or [in other words] our days and years have still much less to say if I reflect on the questions—which you, Sir, have piled up in this connection and to which you are accustomed—with the assurance that in this work of the All-wise no driving wheel gets stuck as often happens with our clocks.

It pleases me very much that you, Sir, deem the shape of the sun's atmosphere worthy of further consideration and even more so that you afford me the hope that [317] you will set up with me the observations pertaining to it and also those concerning our body [regent] in Orion. If memory serves me right, several similar bright spots have been observed in the sky. It is then to be shown what we are going to make of these and of other remarkable features which one has discovered about the fixed stars, now that we have come to the [right] path to view the world-edifice under a form which presumably will not differ too much from the correct one. If the order which we have found in it moves us, just as it is, to admiration though it might not be the true one, then all the more admirable ought to be that order which though still hidden is the real one. I for my part remain with this and will not go further until it becomes scrutinized with observations, and will therefore be found either correct or will be improved. You know, Sir, how intent I am on rigorous proofs. I would very much like to carry out the foregoing scrutiny, namely, to see how far the astronomy with which we have so far luckily succeeded can be rigorously verified. But following our observations this can be done better and more completely. I do not pause, Sir, to describe all my enjoyment which [318] I shall find in all that. But at your arrival I shall not be lacking in words and deeds to manifest the sentiments of joy and unchanging devotion with which I remain, Sir, etc.

193

Herrn Edmund Halley Tafel der Laufbahn, der von ihme berechneten Cometen.

Anni	Nodus ascendens	Inclinatio orbitæ	Perihelium in orbe	Perihelium in eccliptica	Latitudo Perihelii	Distantia Perihelii	Logarithmus Distantiæ	Tempus Perihelii (d. h. ′)
1337	♊ 24° 21′ 0″	32° 11′ 0″	♉ 7° 59′ 0″	♉ 12° 45′ 15″	22° 40′ 30″ B	40666 R	9,609236	Jul. 2. 6. 2
1472	♑ 11° 46′ 20″	5° 20′ 0″	♉ 15° 33′ 30″	♉ 15° 40′ 20″	4° 25′ 50″ A	54273 R	9,734584	Febr. 28. 22. 23
1531	♉ 19° 25′ 0″	17° 56′ 0″	♒ 1° 39′ 0″	♒ 0° 48′ 15″	17° 3′ 5″ B	56700 R	9,753583	Aug. 24. 21. 18½
1532	♊ 20° 27′ 0″	32° 36′ 0″	♋ 21° 7′ 0″	♋ 16° 59′ 40″	15° 17′ 0″ B	50910 D	9,706803	Oct. 19. 22. 1¾
1556	♍ 25° 42′ 0″	32° 6′ 30″	♑ 8° 50′ 0″	♑ 11° 6′ 0″	31° 10′ 20″ B	46390 D	9,666424	Apr. 21. 20. 3
1577	♈ 25° 52′ 0″	74° 32′ 45″	♌ 9° 22′ 0″	♍ 7° 53′ 0″	69° 35′ 20″ A	18342 R	9,263447	Oct. 26. 18. 45
1580	♈ 18° 57′ 20″	64° 40′ 0″	♋ 19° 5′ 50″	♋ 19° 17′ 10″	64° 40′ 0″ B	59628 D	9,775450	Nov. 28. 15. 0
1585	♉ 7° 42′ 30″	6° 4′ 0″	♈ 8° 51′ 0″	♈ 8° 59′ 10″	2° 55′ 25″ A	109358 D	10,038850	Sept. 27. 19. 20
1590	♍ 15° 30′ 40″	29° 40′ 40″	♏ 6° 54′ 30″	♏ 2° 55′ 50″	22° 45′ 50″ A	57661 R	9,760880	Jan. 29. 3. 45
1596	♒ 12° 12′ 30″	55° 12′ 0″	♏ 18° 16′ 0″	♏ 22° 44′ 35″	54° 44′ 30″ B	51293 R	9,710058	Jul. 31. 19. 55
1607	♉ 20° 21′ 0″	17° 2′ 0″	♒ 2° 16′ 0″	♒ 1° 29′ 40″	16° 10′ 5″ B	58680 R	9,768490	Oct. 16. 3. 50
1618	♊ 16° 1′ 0″	37° 34′ 0″	♈ 2° 14′ 0″	♈ 6° 10′ 0″	35° 50′ 0″ A	37975 D	9,579498	Oct. 29. 12. 23
1652	♊ 28° 10′ 0″	79° 28′ 0″	♈ 18° 18′ 40″	♊ 10° 41′ 35″	58° 14′ 0″ A	84750 D	9,928140	Nov. 2. 15. 40
1661	♊ 22° 30′ 30″	32° 35′ 50″	♋ 25° 58′ 40″	♋ 21° 37′ 30″	17° 17′ 0″ B	44851 D	9,651772	Jan. 16. 23. 41
1664	♊ 21° 14′ 0″	21° 18′ 30″	♌ 10° 41′ 25″	♌ 8° 40′ 35″	16° 1′ 50″ A	102575½ R	10,011044	Nov. 24. 11. 52
1665	♏ 18° 2′ 0″	76° 5′ 0″	♊ 11° 54′ 30″	♉ 24° 6′ 35″	23° 8′ 0″ B	10649 R	9,027309	Apr. 14. 5. 15½
1672	♑ 27° 30′ 30″	83° 22′ 10″	♉ 16° 59′ 30″	♋ 9° 26′ 0″	69° 27′ 40″ B	69739 D	9,843476	Febr. 20. 8. 37
1677	♏ 26° 49′ 10″	79° 3′ 15″	♌ 17° 37′ 5″	♋ 16° 21′ 5″	75° 44′ 10″ B	28059 R	9,448072	Apr. 26. 0. 37¼
1680	♑ 2° 2′ 0″	60° 56′ 0″	♐ 22° 39′ 30″	♐ 27° 26′ 50″	8° 11′ 40″ A	612½ D	7,787106	Dec. 8. 0. 6
1682	♉ 21° 16′ 30″	17° 56′ 0″	♒ 2° 52′ 45″	♒ 2° 0′ 30″	16° 59′ 20″ B	58328 R	9,765877	Sept. 4. 7. 39
1683	♍ 23° 23′ 0″	83° 11′ 0″	♊ 25° 29′ 30″	♋ 10° 36′ 55″	82° 52′ 0″ B	56020 R	9,748343	Jul. 3. 2. 50
1684	♐ 28° 15′ 0″	65° 48′ 40″	♏ 28° 52′ 0″	♐ 15° 15′ 25″	26° 35′ 20″ A	96015 D	9,982339	Maj. 29. 10. 16
1686	♓ 20° 34′ 40″	31° 21′ 40″	♊ 17° 0′ 30″	♊ 16° 24′ 0″	31° 17′ 35″ B	32500 D	9,511883	Sept. 6. 14. 33
1698	♐ 27° 44′ 15″	11° 46′ 0″	♑ 0° 51′ 15″	♑ 0° 47′ 20″	0° 38′ 10″ A	69129 R	9,839660	Oct. 8. 16. 57

SELECTED BIBLIOGRAPHY

The following, briefly annotated list is offered as a general orientation on the topic and as a guide to the title of works quoted more frequently in the Notes.

1. Bibliographies on Lambert
    Steck, M., "Materialen zu einer wissenschaftlichen Biographie von Johann Heinrich Lambert," in *Lamberts Schriften zur Perspektive,* edited and introduced by Max Steck (Berlin: Dr. Georg Lüttke Verlag, 1943), pp. 26–35, where 234 titles on Lambert published between 1764 and 1942 are listed in chronological order.
    Steck, M., *Bibliographia Lambertiana. Ein Führer durch das gedruckte und ungedruckte Schriften und den wissenschaftlichen Briefwechsel von Johann Heinrich Lambert 1728-1777* (Hildesheim: Verlag Dr. H. A. Gerstenberg, 1970). This is in part a reprint of the introductory section of Steck's *Lamberts Schriften zur Perspektive,* but it gives a full list of articles published by Lambert in various scientific journals, together with a listing of the items of Lambert's manuscripts in the Gotha Collection (see below), and an essay on the status of research on Lambert. It contains no new entries on the *Cosmologische Briefe.*

2. Writings of Lambert
    In manuscript
    Gotha Collection, or Lambert Nachlass. Following Lambert's death in 1777, his manuscripts and correspondence were purchased by the Berlin Academy and entrusted to Johann Bernoulli (III), director of the Observatory, who published part of Lambert's correspondence (see below, *Briefwechsel*) and part of his essays left in manuscript (see note 6 to the Introduction). Between 1793 and 1799, the whole material together with Bernoulli's own extensive correspondence was purchased by Prince Ernest II of Gotha and transferred to Friedenstein Castle. There the material was bound in 84 volumes, marked 673–756. They were transferred to the Universitätbibliothek Basel in 1938.

    In print
    German-language correspondence with scientific men, edited in five volumes by J. Bernoulli under the title, *Joh. Heinrich Lamberts deutscher gelehrter Briefwechsel* (Berlin: bey dem Herausgeber, 1781–87).
    Lambert-Kaestner correspondence in French, edited by K. Bopp, "J. H. Lamberts and A. G. Kaestners Briefe aus den Gothaer Manuskripten herausgegeben," in *Sitzungsberichte der Heidelberger Akademie der Wissenschaften.* Mathematisch-naturwissenschaftliche Klasse. Jahrgang 1928, 18. Abhandlung (Berlin, 1928), 34pp.
    Lambert-Euler correspondence in French, edited by K. Bopp, "Leonhard Eulers und Johann Heinrich Lamberts Briefwechsel aus den

Manuskripten herausgegeben," in *Abhandlungen der preussischen Akademie der Wissenschaften*. Jahrgang 1924. Physikalisch-mathematische Klasse (Berlin, 1924), 43pp.

Monatsbuch. A diary of 19 quarto pages which contain brief monthly entries from 1752 until 1777 almost exclusively on scientific work done. It was edited by K. Bopp, "Johann Heinrich Lamberts Monatsbuch mit den zugehörigen Kommentaren sowie mit einem Vorwort über den Stand der Lambertforschung," in *Abhandlungen der Könglichen Bayerischen Akademie der Wissenschaften*. Mathematisch-physikalische Klasse. XXVII. Band, 6. Abhandlung (Munich, 1915), 84pp.

Photometria, sive de mensura et gradibus luminis colorum et umbrae (Augsburg: sumptibus viduae Eberhardi Klett, 1760).

Neues Organon, oder Gedanken über die Erforschung und Bezeichnung des Wahren und dessen Unterscheidung von Irrtum und Schein (Leipzig: Johann Wendler, 1764).

Anlage zur Architectonic, oder Theorie des Einfachen und des Ersten in der philosophischen und mathematischen Erkenntniss (Riga: bey J. F. Hartknoch, 1771).

Pyrometrie, oder vom Maasse des Feuers und der Wärme (Berlin: bey Haude und Spener, 1779).

3. Editions of the *Cosmologische Briefe*

Système du monde (À Bouillon: aux dépens de la Société Typographique, 1770). A condensation by B. Merian, published by him anonymously. It is stated only in the Preface that the text is a condensation of the "Lettres Cosmologiques de M. LAMBERT."

Système du monde par M. Lambert; publié par M. Merian (Seconde edition; À Berlin, et se vend à Paris . . . et à Génève . . ., 1784). The Préface is preceded by a two-page-long extract from the review of the first edition in the *Journal encyclopédique* (see note 115 to the Introduction) containing the phrase quoted in that note!

Sistema mira slavnavo Lamberta . . . [Merian's condensation] translated into Russian by Mikhail Rozin (Sankt Petersburg, 1797).

The System of the World, translated from the French by James Jacque, Esq. (London: printed for Vernor and Hood . . . by J. Cundee, Ivy-Lane, 1800).

Lettres cosmologiques sur l'organisation de l'univers, écrites en 1761 par J. H. Lambert. Traduites de l'allemand par Mr. Darquier, . . . publiées et augmentées de remarques par J. M. C. D'Utenhove (Amsterdam: chez Gerard Hulst van Keulen, 1801).

Cosmologische Briefe über die Einrichtung des Weltbaues, im Auszug herausgegeben und mit Anmerkungen versehen von Friedrich Löwenhaupt, pp. 65–108 in F. Löwenhaupt (ed.), *Johann Heinrich Lambert: Leistung und Leben* (Mülhausen [Els.]; Braun & Co., 1943). The total length of excerpts is less than one half of the original. Löwenhaupt's notes are very brief and few.

4. Works on Lambert

Berger, P., "Johann Heinrich Lamberts Bedeutung in der Naturwissen-schaft des 18. Jahrhunderts," *Centaurus,* 6 (1959), pp. 157–254. The *Cosmologische Briefe* is discussed on pp. 167–75, in connection with Lambert's view on teleology, and with emphasis on the identity of thought between Lambert and Kant.

Graf, M., "Johann Heinrich Lambert's Leben." This carefully docu-mented essay of sixty-six pages is the basis of all subsequent accounts of Lambert's life and work, and is the most valuable of the three essays published by D. Huber under the title, *Johann Heinrich Lambert* (see below) in 1829 to commemorate the centenary of Lambert's birth.

Huber, D., *Johann Heinrich Lambert nach seinem Leben und Wirken aus Anlass der zu seinem Andenken begangenen Secularfeier in drei Abhandlungen dargestellt* (Basel: in der Schweighauer'schen Buch-handlung, 1829). In addition to Graf's essay, this volume contains a twenty-page-long bibliography of Lambert's works, an essay by S. Erhardt on Lambert's contributions to theoretical philosophy, and one by Huber on Lambert's attainments in mathematics and physics. The last two pages of this essay are a general summary of the contents of the *Cosmologische Briefe* together with some wholly uncritical remarks as to its significance from the viewpoint of "the developments of the last seventy years in astronomy."

Humm, F., *J. H. Lambert in Chur 1748–1763* (Chur: Calven-Verlag, 1972). A carefully researched work, with useful pages on the *Cosmolo-gische Briefe.*

Jaquel, R., "L'astronome et philosophe mulhousien Jean-Henri Lambert (1728–1777) et les comètes," a series of essays published in the *Bulletin des Professeurs du Lycée d'État de Garçons de Mulhouse* between 1963 and 1968. Special mention should be made of Essay II, "Les conceptions cosmologiques et philosophiques de Lambert relatives aux comètes," 26pp.

Jaquel, R., "Jean–Henri Lambert (1728–1777) et l'astronomie cométaire au XVIIIe siècle," *Comptes-rendus du Quatre-vingt-douzième Con-grès national des sociétés savantes. Strasbourg et Colmar, 1967. Section des sciences. Tome I: Histoire des sciences* (Paris: Bibliothèque Nationale, 1969), pp. 27–56. These essays of Jaquel, together with the ones quoted in notes 9 and 155 to the Introduction, are only the most important among his publications on Lambert and deserve a much wider circulation in view of their very rich and most carefully docu-mented contents.

Löwenhaupt, F., *Johann Heinrich Lambert: Leistung und Leben* (see section above). Although tainted by the political background of its publication, the eight essays comprising the first half of this 4° volume are informative contributions on Lambert the man, the philosopher, and the scientist. Löwenhaupt's appraisal of the *Cosmologische Briefe* (pp. 39–42) lacks a perspective satisfactory from the historical view-point.

Wolf, R., "Joh. Heinrich Lambert von Mühlhausen," in R. Wolf, *Biographien zur Kulturgeschichte der Schweiz. Dritter Cyclus* (Zurich: Orell, Füszli & Co., 1860), pp. 317–56.

5. Other works

Herschel, W., *The Scientific Papers of Sir William Herschel,* edited by J. L. E. Dreyer (London: The Royal Society and the Royal Astronomical Society, 1912).

Hoskin, M. A., *William Herschel and the Construction of the Heavens* (New York: W. W. Norton and Company, 1964).

Jaki, S. L., *The Paradox of Olbers' Paradox* (New York: Herder and Herder, 1969).

Jaki, S. L., *The Milky Way: An Elusive Road for Science* (New York: Science History Publications, 1972).

Kant, I., *Allgemeine Naturgeschichte und Theorie des Himmels* (Königsberg: bey Johann Friedrich Petersen, 1755). There are several modern editions with the original pagination.

Laplace, P. S., *Exposition du système du monde* (Paris: De l'Imprimerie du Cercle-Social, L'An IV de la République Française [1796]).

Newton, I., *Sir Isaac Newton's Mathematical Principles of Natural Philosophy and his System of the World,* Motte's translation revised, and supplied with an historical and explanatory appendix by Florian Cajori (Berkeley: University of California Press, 1962).

NOTES

I NOTES TO THE INTRODUCTION

1. The foregoing account follows the one given by Dieudonné Thiébault in his *Mes souveniers de vingt ans de séjour à Berlin* [etc.] first published in Paris in 1804 and immediately translated into English under the title, *Original Anecdotes of Frederic the Second, King of Prussia, and of his Family, his Court, his Ministers, his Academies, and his Literary Friends; collected during a Familiar Intercourse of Twenty Years with that Prince* (London: Johnson, 1805), vol. II, p. 357.

2. "Éloge de M. Lambert," in *Nouveaux Mémoires de l'Académie Royale des Sciences et Belles-Lettres, Année 1778.* (Berlin: chez George Jacques Decker, 1780), p. 84. The eulogy was delivered on January 29, 1778.

3. As recalled by Johann (III) Bernoulli (1744–1807), astronomer, admirer of Lambert, a colleague of his at the Berlin Academy, and publisher of Lambert's German-language correspondence, in his necrologue, "Précis de la vie de Lambert," printed in French translation in *Lettres cosmologiques,* p. 14.

4. One of those recalling the startling features of Lambert's customary clothing was none other than Johann Elert von Bode (1747–1826). From a distance of well over thirty years those features still stood out vividly in his memory as witnessed by his letter of April 13, 1811, quoted by Graf, "Lambert's Leben," pp. 53 and 62. Bode, twenty years Lambert's junior, owed much of his future career as the leader of German astronomy to his having been brought by Lambert to Berlin to help him with the computations needed for the astronomical ephemerides, which from 1774 on was published by Bode under the title, *Astronomisches Jahrbuch.* Five years before Lambert reached Berlin in 1764, his strange attire and mannerism greatly intrigued in Zurich some youngsters who stopped inconveniencing him only when to their great astonishment the mayor and the city-fathers gathered to greet him. See Humm, *Lambert in Chur,* p. 83.

5. Graf, "Lambert's Leben," p. 34.

6. *Ibid.,* p. 43. Lambert's steadfastness as practicing Christian should be kept in mind when evaluating the report by C. H. Müller, Lambert's colleague at the Berlin Academy, that with respect to basic Christian dogmas Lambert's thought shifted toward rationalism during the last six years of his life. See Müller's "Bemerkungen über Lamberts Character," in *Joh. Heinrich Lamberts logische und philosophische Abhandlungen,* edited by Joh. Bernoulli, vol. II, (Berlin: bey dem Herausgeber, 1787), p. 368.

7. Thiébault, *Original Anecdotes* (see note 1 above), vol. II, p. 358.

8. As reported in the autobiography of Johann Georg Sulzer (1720–1779), member of the Berlin Academy from 1750, and a main promoter of Lambert's cause at Frederick's court. See Graf, "Lambert's Leben," p. 16.

9. The question of Lambert's nationality was made particularly complicated by late-nineteenth-century chauvinism which claimed Lambert for Switzerland as well as for France and Germany. For a carefully documented discussion, see R. Jaquel, "Le problème nuancé de la nationalité du 'Leibniz Alsacien' Jean-Henri Lambert (1728–1777)," in *Bulletin du Musée Historique de Mulhouse,* 79–81 (1971–73), pp. 81–109. At the start of his otherwise very useful and informative essay, "Lambert von Mühlhausen" (p. 317), R. Wolf, astronomer of Zurich, who made name for

himself with his publications in the history of astronomy, stated in 1860 that "Lambert consistently considered himself a Swiss, and until he had no other titles he was designated by his contemporaries as Mülhousino-Helvetus."

10. Formey, "Éloge" (see note 2 above), p. 74.

11. On these details, see Graf, "Lambert's Leben," pp. 3–6.

12. The essay appeared as Appendice II, in Chéseaux's *Traité de la comète*, pp. 223–29. For the full text of that essay, see my *The Paradox of Olbers' Paradox*, pp. 253–55.

13. *Photometria*, pp. 504–11. See also my *The Paradox of Olbers' Paradox*, pp. 123–25.

14. *Briefwechsel*, vol. II, pp. 7–11.

15. *Ibid.*, pp. 8–9. The books referred to are Christian Friedrich Wolff's *Vernünftige Gedancken von den Kräfften des menschlichen Verstandes und ihrem richtigen Gebrauche*, Nicolas Malebranche's *De la recherche de la vérité*, and John Locke's *An Essay concerning Humane Understanding*.

16. *Briefwechsel*, vol. II, p. 9. Lambert refers to Samuel Pufendorf's *De officio hominis et civis* (1675).

17. *Ibid.* Charles Rollin (1661–1741) was Rector of the University of Paris. His writings related to history, theology, rhetorics, and education. The work quoted by Lambert was first published in 1726–28 in four volumes under the title: *De la manière d'enseigner et d'étudier les belles-lettres, par rapport à l'esprit et au coeur.*

18. *Briefwechsel*, vol. II, p. 10.

19. Lambert's last, relatively brief stay with the Salis family took place in 1763. See Humm, *Lambert in Chur*, pp. 107–15.

20. The Monatsbuch was written in Latin.

21. "Tentamen de vi caloris, qua corpora dilatat, eiusque dimensione," in *Acta Helvetica, physico-mathematico-anatomico-botanico-medica* (Basel) 2 (1755), pp. 172–242.

22. Published in 1758. In the three articles Lambert dealt with the theory of balance (pp. 13–22), with problems of convergence in series (128–68), and with meteorological observations made in Chur (pp. 321–65). The fourth volume of the *Acta* (1760) carried Lambert's analysis of the variations of barometric pressure due to the position of the moon (pp. 315–36).

23. *Les propriétés remarquables de la route de la lumière* [etc.], (The Hague: chez H. Scheurleer, F. Z. & Compagnie, 1758).

24. This small 8° book of 116 pages contained thirty-three theorems, the material of which was methodically distributed into experiments, observations, demonstrations, lemmas, corollaries, problems, and solutions. Clearly, the young scholar aimed at exactness and rigor, and possibly wished to imitate the structure of Newton's *Principia* and other major scientific works. Lambert referred to Bouguer's work as "un très bel Essai" on p. 5 of the Avant-propos. The work in question was the *Essai d'optique sur la gradation de la lumière* (Paris: chez Claude Jombert, 1729) in which the second half of Section II deals with "the proportion according to which light decreases in passing through [various] media." See especially pp. 44–48. For the second edition of this work, see note XIV-6.

25. *Die Freye Perspective* (Zurich: Heidegger, 1759), or as its subtitle states, "a guide how to work out any visual sketch from individual details and without a general plan." For its critical and annotated edition, see Steck, *Lamberts Schriften zur Perspektive.*

26. See the Anmerkungen by E. Anding to his translation, *Lambert's Photometrie* (Ostwald's Klassiker der exakten Wissenschaften, Nr. 31–33; Leipzig: Wilhelm Engelmann, 1892), vol. III, pp. 51–172.

27. *Insigniores orbitae cometarum proprietates* (Augsburg: E. Klett, 1761). The outstanding features of this book, its place in the history of cometary studies, and its relevance for the role played by comets in Lambert's cosmological thought are presented with meticulous care in R. Jaquel's essay, "Lambert et l'astronomie cométaire."

28. The essay, "Über die Methode die Metaphysik, Theologie und Moral richtiger zu beweisen," was published by Karl Bopp in 1918 in *Kantstudien*. Ergänzungsheft Nr. 42.

29. Kant's essay is available in English translation under the title, "Enquiry concerning the Clarity of the Principles of Natural Theology and Ethics," in *Kant: Selected Pre-critical Writings and Correspondence with Beck,* translated and introduced by G. B. Kerferd and D. E. Walford, with a contribution by P. G. Lucas (Manchester: University Press, 1968), pp. 3–35. In his letter of Dec. 31, 1765, to Lambert, Kant spoke of him as "the first genius in Germany who is capable of achieving an important and lasting improvement in the kind of investigations which kept especially busy" Kant as well. See *Briefwechsel,* vol. I, p. 340.

30. Lambert's second major philosophical work, *Anlage zur Architectonic,* "or theory of the simple and primary in philosophical and mathematical knowledge," was published in 1771. These two works appeared so "Kantian" to Kant that he planned at one point to dedicate his *Critique der reinen Vernunft* to Lambert. For the draft of that dedication, see B. Erdmann (ed.), *Reflexionen Kants zur Kritik der reinen Vernunft* (Leipzig: Fues's Verlag, 1884), pp. 1–2.

31. Much of the text of the Antrittsrede was printed in *Histoire de l'Académie Royale des Sciences et Belles-Lettres. Année MDCCLXV* (Berlin: chez Haude et Spener, 1767), pp. 506–14.

32. A remark of Thiébault who often accompanied Lambert for a walk and tried to deviate him from his train of thought but in vain: "He was truly a machine to grind dissertations, but a perfect machine." *Original Anecdotes,* vol. II, p. 360.

33. Doubts about Lambert's priority in reaching these conclusions were definitively laid to rest by A. Pringsheim, "Über die erste Beweise der Irrationalität von e und π," in *Sitzungsberichte der Bayerischen Akademie der Wissenschaften,* Math.-phys. Klasse, Band XXVIII (1898), pp. 325–37. See also the paper, "Johann Heinrich Lambert," read by W. Lorey on the bicentenary of Lamberth's birth in the Mathematical Society of Berlin. *Sitzungsberichte der Berliner Mathematischen Gesellschaft,* 18 (1928), pp. 2–27. See especially, pp. 19–20.

34. Graf, "Lambert's Leben," p. 38.

35. The letter (*Briefwechsel,* vol. II, pp. 46–50) written to an anonymous friend in Zurich, first deals with the shortsighted political trends in Switzerland and contains sharp remarks on Albert Haller, the famed physiologist, who after a brilliant career in Göttingen returned to his country, and following a brief involvement in politics voiced his disillusion with public affairs. Although, as Lambert put it, quite a few Swiss seemed to prefer "to be bailiffs under French rule than to be free citizens in their own land," he still held out the hope that good sense and patriotism would prevail. But the more distant future appeared to him rather gloomy: "At any rate, according to my opinion, it follows from all actual circumstances that the

developments point not only for Switzerland but also for the whole of Europe to an impending great revolution which is, of course, not completely ripe for an outbreak and will perhaps take place only in the next century. For the time being there is calm and peace in the most populated areas of Europe but not balance [of power]. With the sole exception of the House of Bourbon [France] all are either neutral or for Russia. The latter can come unopposed into possession of the entire European [part of] Turkey. No one has any claim on Greece proper, or the seat of the expired oriental [Byzantine] empire. Austria can demand back the stolen parts of Hungary; Venice that of Dalmatia and some islands. In return for her help rendered on the sea England will probably demand some islands and ports useful for commerce; and when things get that far I do not know what France can do except to counsel the Turks once more against war. The next few years must provide more light here. England herself will have to come to the point of putting her American colonies on the status of Ireland, giving them a parliament and a viceroy, or turning them into an actual part of Great Britain; and this part could, because it is so heavily populated that its inhabitants double in every ten years, become the chief part [of Great Britain] in the future. This can happen if things go well for England; in the opposite case the colonies make themselves sovereign on their own, and can attain a very great power. About Russia it is assumed that because of her vastness she will never muster her forces, or perhaps will fall apart because of internal disturbances. There is no evidence yet for either. Russia is in the position that Macedonia was at the time of Philip, when the smaller republics kept bickering about the balance [of power], as it has happened until now in Europe. Only time will tell therefore whether a conquering spirit, like Alexander [the Great], will ascend the Russian throne. Then even Switzerland can be involved in greater upheavals. We shall hope meanwhile that this is still far away, or rather that this will never take place." That this letter was written on the 1st of May, 1770, may add something to its prophetic ring. A few and somewhat disconnected phrases from the section quoted above are given in Graf's "Lambert's Leben" (p. 52), and in Löwenhaupt's *Lambert: Leistung und Leben* (p. 55).

36.   The scarcity of the original can be gathered from the mere twenty-two copies that Steck was able to locate in major libraries in Germany and Austria prior to 1943. See Steck, *Lamberts Schriften zur Perspektive,* p. 99. The *National Union Catalogue of pre-1956 Imprints* lists ten American libraries with a copy.

37.   Monatsbuch, p. 22.

38.   In a letter written on November 13, 1765. See *Briefwechsel,* vol. I, p. 336. For longer sections of this letter in English translation, see my *The Milky Way,* pp. 199–200.

39.   *Briefwechsel,* vol. I, p. 418. Boeckmann was professor of mathematics and physics in Karlsruhe.

40.   In his *Discours sur les différentes figures des astres,* published in 1742, P. L. M. de Maupertuis attributed the different appearances of "nebulous stars" to the fact that they were "flattened ovals" whose equatorial planes had a great variety of positions with respect to us. See *Oeuvres de Maupertuis* (new ed.; Lyons: Jean-Marie Bruyset, 1756), p. 143.

41.   During the 1750's there had been several communications to the Académie des Sciences in Paris on "nebulous" stars. The first to appear in print was Guillaume Le Gentil's "Mémoire sur une étoile nébuleuse nouvellement découverte à côté de celle qui est au dessus de la ceinture d'Andromède," in *Mémoires de mathématiques et de physique.* Présentés à l'Académie Royale des Sciences par divers Savans et lus

dans ses Assemblées. Tome Second (Paris: De l'Imprimerie Royale, 1755), pp. 137–44. Le Gentil stressed the changeability of the appearance of nebulae and referred (p. 142) to the discussion by J. J. D. de Mairan of the changes observed in the great nebula in Orion. De Mairan's discussion was easily available in his massive monograph on the aurora borealis, *Traité physique et historique de l'aurore boréale* (2d ed; Paris: De l'Imprimerie Royale, 1754), pp. 261–62. This second edition was almost twice as large as the first (1733), and both largely consisted of de Mairan's communications to the Académie. The general report, "Sur les nébuleuses," presented most likely by Le Gentil to the Académie in 1759, mentioned some changes in that nebula, but it was printed only in 1765, in *Mémoires de l'Académie Royale des Sciences. Année 1759.* (Paris: De l'Imprimerie Royale, 1765), pp. 183–85. The same volume carried also Le Gentil's "Remarques sur les étoiles nébuleuses" (pp. 453–71), in which Le Gentil's own observations were introduced with a history of speculations on "nebulous stars." Moreover, Le Gentil offered four engravings, one by Huygens, one by Picard, and two by himself, which suggested a variation of the hue and shape of the great nebula in Orion.

42. The return of Halley's comet was expected to take place according to Halley's prediction some time in 1758, but was first spotted only on Christmas eve that year. It reached its maximum visibility during the early spring of 1759. See A. Armitage, *Edmond Halley* (London: Nelson, 1966), pp. 165–66. It was the first predicted return of a comet to be observed, and created a brief dispute because some had taken its slight delay as an evidence of the unreliability of the studies of Halley and others on comets.

43. See his letter of Feb. 3, 1761, to Gessner, *Briefwechsel,* vol. II, p. 186. His appointment came in December 1759. In April 1761 the organization of the class of physical sciences was entrusted to him. A year later Lambert changed his status to that of a corresponding member. The Academy kept receiving from him a steady flow of papers for which he received an annual stipend of 200 guldens. See Humm, *Lambert in Chur.* p. 110.

44. See note 27 above.

45. Monatsbuch, p. 22.

46. *Briefwechsel,* vol. I, 418–19. See also p. [299] of translation.

47. Monatsbuch, p. 22.

48. See Preface, p. [XVI].

49. Monatsbuch, p. 23.

50. *Ibid.,* p. 23. That the printing was completed in two months' time is clear from Lambert's letters to be quoted shortly.

51. The sequence of pages from p. XVIII on is as follows: XXI, XX, XXV, XXII, XIX, XXIV, XXVII, XXVI, XXIII, XXVIII.

52. Explanatory notes that might have been invited by some details in this résumé are given in connection with their occurrence in the text of the Letters.

53. See Selected Bibliography. Shortly after its publication the publisher went bankrupt and his holdings were impounded by fiscal authorities with the result that only a few copies of Kant's work reached the public.

54. The notion of a general similarity between the two works was first elaborated in some detail during the latter part of the nineteenth century for reasons that will be discussed below. See note 190 to this Introduction.

55. In *Kant's Cosmogony,* the English translation by W. Hastie of Kant's *Allgemeine Naturgeschichte,* the Third Part is omitted possibly because of Hastie's concern that its perusal might arouse suspicion even in the average reader about the scientific merits of much of the first two Parts. Yet the omission was quite an un-Kantian procedure, as Kant did not disavow the Third Part when in the 1790's, largely because of Herschel's discoveries, Kant was raised by his admirers in Germany to the status of Herschel's forerunner and his *Allgemeine Naturgeschichte* was reprinted several times within a few years. Needless to say, the three reprints of Hastie's translation that appeared during the last ten years do not contain the Third Part. It will appear in my translation in the Proceedings of the International Colloquium III, Cosmology-History-Theology, November 5–8, 1974, University of Denver, to be published by Plenum Press.

56. See especially his celebrated "Deutsche Physik," the popular name for his *Vernünftige Gedancken von den Absichten der natürlichen Dinge,* which came out in five editions between 1724 and 1752.

57. Lambert, who in his letter of November 13, 1765 to Kant (see note 38 above) recalled that he had heard of Wright's work (Lambert merely referred to an "Englishman" and did not quote the title, *An Original Theory and New Hypothesis of the Universe,* published in London in 1750), most likely had never seen a copy of it. On Wright's conception of the Milky Way and on the persistent misrepresentations of it in the astronomical literature, see Chapter VI, "Wright's Wrong," in my *The Milky Way.* The uncounted number of galaxies, which Wright postulated, served the purpose of providing a great variety of habitats in the perspective of supernatural renumeration.

58. Swedenborg's *Principia rerum naturalium,* the first of the three volumes of his *Opera philosphica et mineralia,* was published in Dresden in 1734. It is a classic case of infatuation with a particular part of physical science (magnetism in Swedenborg's case), which is uncritically foisted on small and big details in the universe as the supreme explanatory device. Swedenborg's concept of the Milky Way is analyzed in my *The Milky Way,* pp. 168–71.

59. See, for instance, Jean Bernoulli's essay, "Nouvelles pensées sur le système de M. Descartes et la manière d'en déduire les orbits et les aphélies des planètes" (1730) in *Johannis Bernoulli . . . opera omnia* (Lausanne: M. M. Bousquet et Sociorum, 1742), vol. III, pp. 133–73. There is a long list of articles under the entry "Tourbillons" in the *Table générale des matières continues dans l'Histoire & dans les Mémoires de l'Académie Royale des Sciences depuis l'Année 1741 jusqu'à l'Année 1750 inclusivement,* compiled by M. Demours (Paris: Par la Compagnie des Libraires, 1758).

60. See, for instance, the first of the five volumes of *Physique du monde* by Étienne-Claude, baron de Marivetz, and M. Goussier (Paris: Imprimerie de Quillau, 1780–87).

61. Buffon's theory first saw print with the publication in September 1749 of the first three volumes of his *Histoire naturelle.* A detailed discussion of his theory together with an account of the reactions to it is given in my *Planets and Planetarians: A History of Theories of the Origin of Planetary Systems,* to be published shortly.

62. See note 13 above.

63. Both in his Boyle Lectures and in his correspondence with Newton. For details, see my *Paradox of Olber's Paradox,* pp. 61–65, and *The Milky Way,* pp. 152–55.

64. For the text of Halley's discussion and for its analysis, see my *Paradox of Olbers' Paradox*, pp. 249-52 and 76-83.

65. It was that interest which gave rise between 1740 and 1783 to the massive monographs on comets by Struyck, Dionis du Séjour, and Pingré.

66. Lambert-Kaestner, p. 15.

67. Gotha Collection 732*** ff. 44v–45r.

68. *Ibid.*, f. 45r.

69. *Ibid.*

70. Stück 22, pp. 211–13.

71. *Ibid.*, p. 213.

72. The letter, written in French, is dated Jan 24, 1764. See Lambert-Kaestner, pp. 15–16.

73. Lambert-Euler, pp. 18–19.

74. *Ibid.*, pp. 22–25. The absence of any reference to the *Cosmologische Briefe* should seem all the more surprising because several topics discussed by Lambert were most germane to its contents.

75. See especially Lambert's letter of July 24, 1763, Gotha Collection 745 f. 99r.

76. Gotha Collection 732*** f. 45v.

77. Gotha Collection 732*** f. 49r. Several sections of this long letter of six manuscript pages (46v–49r) written in French are quoted by Humm, *Lambert in Chur*, p. 100.

78. Gotha Collection 707 f. 477r.

79. Johann Jakob Bodmer (1698–1783) of Zurich was a forerunner of Romantic literature. In his epic poem, *Noah: ein Heldengedicht* (1750), Bodmer speaks of Sipha, an astronomer and father of three girls who married the three sons of Noah.

80. Gotha Collection 707 f. 477r. This section too is quoted by Humm (*Lambert in Chur*, pp. 98–99) in the original French.

81. A claim of Christoph J. Scriba made in his article "Lambert" in the *Dictionary of Scientific Biography*, vol. VII, (New York: Charles Scribner's Sons, 1973), p. 598. Humm's claim (*Lambert in Chur*, p. 97) that Lambert "erhielt durch seine Kosmologie . . . von denkenden Zeitgenossen viel Anerkennung" is even more uncritical because it was precisely Lambert's "denkenden" scientific contemporaries who ignored his cosmology.

82. *Briefwechsel*, vol. I, p. 371.

83. *Ibid.*, pp. 372–73.

84. *Ibid.*, pp. 373–74.

85. *Ibid.*, p. 374.

86. *Ibid.*, p. 375.

87. *Der einzig mögliche Beweisgrund zu einer Demonstration des Daseins Gottes* in Kant's *Gesammelte Werke*, edited by the Prussian Academy of Sciences, vol. II, (Berlin: Georg Reimer, 1905). For the summary in question, see Abt. II, Betrachtung "Kosmogonie," pp. 137–50.

88. *Ibid.*, pp. 68–69.

89. The second, third, and fourth printings came in 1770, 1783, and 1794.

90. *Le Nouvelliste Suisse, historique, politique, litéraire et amusant,* Décembre 1763, (Neuchâtel: de l'Imprimerie des Editeurs, 1763), pp. 665–67. This periodical was part of the *Journal helvétique* and was often referred to as such.

91. *Le Nouvelliste Suisse,* Janvier 1764, pp. 3–30.

92. *Le Nouvelliste Suisse,* Février 1764, pp. 115–30 and 130–50. The translations of both Letters were accompanied by lengthy notes, some of which will be referred to in the Notes to the *Cosmological Letters.*

93. Tome IV for that year; Deuxième Partie pp. 3–21, and Troisième Partie pp. 10–33.

94. Troisième Partie, pp. 13–14, where a good summary is given of Lambert's explanation of the Milky Way. This is another evidence of the curiously imperceptible way in which the true notion of the Milky Way entered general consciousness between 1750 and 1785, that is, between the publication of Wright's *Original Theory* and the two great memoirs of Herschel on the construction of the heavens. For further details, see Chapters VI and VII in my *The Milky Way.*

95. Troisième Partie, pp. 32–33.

96. Gotha Collection 707 f. 450r–v.

97. In a letter of Jan. 16, 1767. Gotha Collection 707 f. 451r.

98. *Briefwechsel,* vol. I, p. 389.

99. *Briefwechsel,* vol. I, pp. 335–37.

100. The principal source of information on Le Sage is the work by Pierre Prévost, *Notice sur la vie et des écrits de George-Louis Le Sage . . . rédigée d'après ses notes* (Geneva: chez J. J. Paschoud, 1805), in which many excerpts are given from the correspondence between Lambert and Le Sage, but not the ones with reference to the *Cosmologische Briefe,* an uncanny indication of the unimportance which Prévost seemed to attribute to it.

101. Gotha Collection 745 f. 29v.

102. *Ibid.,* ff. 73v–74r. Parts of that letter are also quoted in the original French by Humm, *Lambert in Chur,* p. 95. The manuscript of this teleological treatise on terrestrial processes consists of 26 quarto pages and is registered under the title "Briefe über den Optimismus." See Monatsbuch, p. 47.

103. Gotha Collection 745 ff. 75r–79v.

104. *Ibid.,* f. 75v. This declaration was preceded by a reference to the impossibility of the gradual formation of a planet on the basis of Le Sage's theory: "If you require centuries not for the formation of a planet but for that of a cubic inch of matter by the accumulation of particles which attach to one another, more than an eternity would be necessary to produce entire planets." *Ibid.,* f. 75r.

105. *Ibid.,* ff. 78v–79r. Lambert concluded: "Actually one now begins in France to introduce inversions, and if this is not against the spirit of the language and if habit does not oppose it too much, the French language will be less stilted than it has been for almost a century." A curious remark by one whose style could on occasion be extremely stilted.

106. *Ibid.,* f. 82v.

107. *Ibid.* Here Lambert gives the following summary of the work of Kant, "who considers Sirius as the central body of the world, who considers the whole assemblage of fixed stars as located in the plane of the Milky Way, but who speaks only of 28 or

30 comets." Lambert's infatuation with comets clearly blinded him to the real shortcomings in Kant's work.

108. *Ibid.*, ff. 75v–76r. The statement deserves to be quoted in full: "It seems to me rather certain that the occult qualities have been a bit too much decried since Descartes. One should only blame the abuse of that term by the Scholastics. For this term was introduced only to complete the classification of the different ways of knowing things. Hence the division of qualities into sensible and non-sensible, or occult. One does injustice to the Scholastics in several other respects as well, and one very often decried a [Scholastic] system [of thought] for the sole reason that one was too lazy to study it thoroughly, or because one wanted to earn the title of conqueror, of destructor, or chief of a new sect. Very often one changed but names. It is true, however, that in physics the Scholastics fell a bit too behind, because of a lack of experimenting and because the laws of motion were unknown to them. But in other parts of philosophy we are hardly more advanced, and I finally succeeded in showing that as long as the philosopher does not carry his researches to the point where the geometer can apply his calculus, the philosopher can positively rest assured that he is still in an absolute confusion. I have also remarked that everywhere where some knowledge has been submitted to calculus, it completely changed complexion, because one had to look at it in a wholly different manner." The passage is a perfect example of the inconsistencies of judgment of a philosopher who is tossed back and forth between his common sense and his addiction to a purely quantitative and logical type of rationalism.

109. In a letter of Dec. 11, 1768; *Briefwechsel*, vol. I, p. 316.

110. *Ibid.*, Lambert obviously had in mind Bonnet's *Contemplation de la nature*, an enormous success from its first publication in 1764.

111. *Ibid.*

112. *Briefwechsel*, vol. II, p. 57.

113. *Ibid.*, pp. 58–59, in a note by the editor, Bernoulli, who identifies the student as Mr. Pfleiderer.

114. See Selected Bibliography. This summary consisted of two parts of which the first covered the first eight Letters. The total length was 188 small 8° pages. Merian (1723–1807), professor at the University of Basel, was brought by Maupertuis to the Berlin Academy. A prolific author, Merian became a director there in 1770.

115. For the review, see pp. 331–49; for quotations, see pp. 332–33. About half of the review is a series of quotations from Merian's condensation. While the principles of Lambert's cosmology are clearly listed, no reference is given to the dark central bodies, or regents. It was perhaps the complete lack of publicity in France about the original that prompted the reviewer to remark: "Mr. Merian declares (about which he could have just as well kept silent) that he took from Lambert's *Cosmological Letters* the plan and divisions of that system [presented by him]" (p. 332).

116. See Toaldo's letters of March 17 and Aug. 15, 1771, and of Aug. 10, 1772, to Lambert; Gotha Collection 725 ff. 117r, 120v, and 129r. Toaldo (1719–1798), a Catholic priest, was professor of physics at the University of Padua and well known for his books on meteorology, for his translation into Italian of Lalande's *Astronomie*, and for his edition of Galileo's works.

117. See Lambert's letters of Feb. 8 and April 20, 1771, to Toaldo; Gotha Collection 707 ff. 541v and 543r.

118. Gotha Collection 745 f. 89r: "J'ai demandé à Mr. Preverdil ce qu'on pense à Génève du Journal Encylopédique au sujet de l'extrait du Système du Monde de Mr. Merian. Il me repondoit en deux mots, que le journaliste est un fou. Je n'en demande pas d'avantage, quoiqu'une partie semble tomber sur Mr. Merian lui-même. Ce seroit à lui de redresser les bevues."

119. This letter of Nov. 22, 1771, of Bonnet is quoted by Wolf, "Lambert von Mühl-hausen," p. 333. A month later Bonnet, in a letter of Dec. 24, to Haller again, stressed that the universal harmony of all was expressed in Lambert's cosmology incomparably more beautifully than anywhere else. *Ibid.*

120. Quoted by Wolf, "Lambert von Mühlhausen," p. 335.

121. In a letter of Feb. 20, 1773; *Briefwechsel,* vol. I, p. 415. It was in reply to this letter that Lambert spoke of the origins of his writing the *Cosmologische Briefe.* See note 39 above.

122. For quotations, see pp. 9 and 37 in Lalande's forty-page-long brochure, *Réflexions sur les comètes qui peuvent approcher de la Terre* (Paris: chez Gibert, 1773), which was to be a part of a longer memoir of Lalande to the Académie. On the basis of those rumors Lalande was reported in the May 7, 1773, issue of the *Gazette de France,* that he had predicted the collision of the earth with a comet soon to appear.

123. *Essai sur les comètes en général, et particulièrement sur celles qui peuvent approcher de la Terre* (Paris: Valade, 1775). In the "Discours préliminaire" (pp. i–xxiii) Whiston, Maupertuis, Lalande, and others are mentioned but not Lambert, whose method of computing cometary orbits was ignored in a book almost wholly devoted to their mathematical analysis.

124. *Ibid.,* pp. 337.

125. *Ibid.,* Section IX, pp. 197–210.

126. *Ibid.,* Section VIII, pp. 184–216.

127. *Ibid.,* p. 343.

128. See Lambert's letter of March 27, 1775, to Lalande; Gotha Collection 706 f. 395r.

129. Letter of Feb. 21, 1775, to Maupertuis; *ibid.,* f. 411r.

130. Those 234 letters were written by Euler between April 19, 1760 and May 18, 1762 as an aid of instruction to a relative of Frederick the Great, but were published in 1768, three years after Euler had returned from Berlin to St. Petersburg. The *Lettres à une princesse d'Allemagne sur divers sujets de physique et de philosophie,* an instant success, could have readily prompted a comment or two on the part of Lambert, the cosmologist, as Euler took the view in Lettre CCXXXI that with the removal of the atmosphere only the sun's disk and the point-images of stars would be visible during daytime and "the entire rest of the sky would be as dark as night." (See edition at Mietau and Leipzig, chez Steidel et Compagnie, 1774, vol. III, p. 429). Such could hardly be an acceptable conclusion in view of the immensity, let alone infinity, of stars, a problem that Lambert carefully discussed in his *Photometria.* See note 13 above.

131. In a letter of Feb. 10, 1775. Gotha Collection 707 f. 103r. Merian's summation of Lambert's cosmological ideas "enthralled me," Trembley wrote, "to the point of making the resolve to read at any price all that issued from your pen. I have therefore ordered all your German works to be sent to Zurich and I have set myself to studying that language." A few lines later, with a reference to the *Neues Organon*

and the *Architectonic,* Trembley admitted that Lambert's works were "too high" for him (f. 103v). There are no references to the *Cosmologische Briefe* in the four letters exchanged between Lambert and Fontana between May 5, 1775 and Aug. 20, 1776, although several other works of Lambert were discussed (Gotha Collection 705 ff. 71r–76v).

132. *Briefwechsel,* vol. I, pp. 425–28.

133. *Ibid.,* pp. 428–29.

134. First published in 1768, under the title, *Deutliche Anleitung zur Kenntniss des gestirnten Himmels* in Hamburg. The work now owes its fame to its second edition, published in 1772, in which Bode discussed the regularity of planetary distances, subsequently known as the Titius-Bode Law. See my article, "The Early History of the Titius-Bode Law," *American Journal of Physics,* 40 (1972), pp. 1014–23.

135. The three and a half pages of the "Nachschrift" at the end of the volume are unnumbered; *Astronomisches Jahrbuch oder Ephemeriden für das Jahr 1780* (bey George Jacob Decker, 1777).

136. *Kurzgefasste Erläuterung der Sternkunde und den* [sic] *dazu gehörigen Wissenschaften* (Berlin: bey Christian Friedrich Himburg, 1778), vol. II, p. 512: "Kant in his *Allgemeine Naturgeschichte* and similarly Lambert in his *Cosmologische Briefe* have, as true philosophers, discussed this subject [the Milky Way and the construction of the heavens] with speculations worthy of the greatness of God." In the second considerably enlarged edition (1793) the only significant change in the same sentence was the adding of Herschel's name (vol. II, p. 728).

137. See his "Éloge" (note 2 above), p. 85.

138. See the French translation prefixed to the *Lettres cosmologiques,* pp. 9–10.

139. See its reprint in Steck, *Lamberts Schriften zur Perspektive,* p. 17.

140. For an English translation of the relevant part of that letter, see my *The Milky Way,* p. 236.

141. *Ibid.*

142. *Cométographie ou Traité historique et théorique des comètes* (Paris: de l'Imprimerie Royale, 1783–84).

143. See especially vol. II, pp. 442–52.

144. *Ibid.,* pp. 110–18.

145. *Ibid.,* pp. 188–216.

146. *Ibid.,* p. 117. According to Pingré the number of comets could not be determined.

147. *Ibid.,* pp. 171–73. Planets detached by a comet from the sun could only go into a very elongated orbit of which the point of collision remained a part.

148. *Ibid.,* p. 116.

149. See note 119 above.

150. *Oeuvres d'histoire naturelle et de philosophie de Ch. Bonnet,* vol. VII, (Neuchâtel: De l'Imprimerie de Samuel Fauche, 1781), pp. 33–35. In another, large 8° edition of the *Oeuvres* published simultaneously, see vol. IV, pp. 22–23.

151. *Ibid.,* vol. VII, p. 33.

152. *Ibid.,* p. 34.

153. *Ibid.*

154. See Selected Bibliography.

155. See Selected Bibliography. An informative account of the circumstances of this translation and of the translator's life and approach to his project is given in R. Jaquel, "L'astronome toulousain Darquier (1718–1802) et le cosmologue mulhousien Jean-Henri Lambert (1728–1777)," in *Comptes rendus du Quatre-vingt-seizième Congrès National des Sociétés Savantes. Toulouse 1971*. Section des Sciences. Tome I. Histoire des Sciences (Paris: Bibliothèque Nationale, 1974), pp. 31–46. Utenhove (1773–1836) was an amateur but competent astronomer, whose numerous notes put Lambert's work in the context of the astronomical discoveries and speculations of the four decades following its publication. Because of the historical value of his notes, they will be constantly recalled in the Notes to the text.

156. I am indebted for this information to a private communication from Mr. Roger Jaquel, professeur honoraire au Lycée Albert-Schweitzer à Mulhouse, whose meticulous researches on Lambert deserve the attention not only of Lambert scholars but of all specialists of eighteenth-century astronomy.

157. See Selected Bibliography.

158. It appeared in M. K. Munitz (ed.), *Theories of the Universe: From Babylonian Myth to Modern Science* (New York: The Free Press, 1957), pp. 250–63. The note of reference on p. 250 states: "From John Henry Lambert, *The System of the World,* translated ("digested") from the French by James Jacque, London, 1800, Part II, Chapters iv, vii, viii, ix." In the general introduction to the section, "The Copernican Revolution and its Aftermath," Munitz states as a final remark on Kant's cosmogony: "A similar speculative cosmology, making use of the notions of Newtonian mechanics and the idea of a hierarchy of celestial systems culminating in a single infinite universe was worked out by Johann Lambert, a contemporary of Kant, in his *Cosmological Letters,* along lines that are in some respects strikingly similar to those of Kant" (p. 145). The idea of a "single infinite universe" is certainly not advocated by Lambert in the *Cosmological Letters!*

159. *Abhandlung über die leichteste und bequemste Methode die Bahn eines Cometen aus einigen Beobachtungen zu berechnen* (Weimar: im Verlage des Industrie-Comptoirs, 1797). For the list of the elements of 89 cometary orbits, see Appendix, pp. 36–50.

160. See the Vorrede by Zach, p. xxix.

161. *Aphroditographische Fragmente zu genauern Kenntniss des Planeten Venus* [etc.] (Helmstedt: C. G. Fleckeisen, 1796).

162. *Ibid.,* pp. 245–50, and Taf. II.

163. *Ibid.,* pp. 193–94.

164. One of them was J. D. Cassini, director (1784–93) of the Paris Observatory, who, however, hastened to assure the world of science that Herschel's reports were trustworthy. See my *The Milky Way,* p. 231.

165. The best account of that procedure is in M. Hoskin, *William Herschel and the Construction of the Heavens* (New York: W. W. Norton Company, 1964), pp. 88–92.

166. As Hoskin pointed out, Herschel owned a copy of Wright's *Original Theory* and some ideas peculiar to Wright turn up here and there in Herschel's papers. See Hoskin's Introduction (p. xxviii) to his facsimile edition of Wright's work (London: Macdonald, 1971).

167. A list of such works is given by Hoskin, *ibid.*, p. xvi.

168. A documentation of points concerning Herschel can readily be had in Hoskin, *William Herschel and the Construction of the Heavens.*

169. "On the Direction and Velocity of the Motion of the Sun, and Solar System" (1805), in *Scientific Papers,* vol. II p. 318. It is important to note that Herschel's remark was far more than a facile criticism. Lambert's cosmological speculations, which he knew only from Merian's *Système du monde* (he referred to it in the footnote as a work by Lambert), seemed to him to exemplify a frame of mind in which factual data, or their absence, were readily overlooked because of eagerness to build systems. The primacy of factual data was in Herschel's eyes a principle not to be tampered with even if it made the system-building more difficult. A case in point was the question of the sun's motion. If established, it could, as Herschel noted, "reveal so many concealed real motions, that we shall have a greater sum of them than it would be necessary to admit, if the sun were at rest." Since simplicity was an unquestionable advantage, a fact undermining it needed all the verification that could possibly by mustered. But to consider with Lambert the motion of the sun, of the planets, and of their satellites as constituting a set of links which made well-nigh real the existence of many other similar links, was not a procedure for Herschel's liking. While admitting some of its plausibility, he fully saw its dangers: "Those who like to indulge in fanciful reviews of the heavens, might easily build a system upon hypotheses not altogether without some plausibility in their favour. Accordingly we find that Mr. Lambert, in a work which is full of the most fantastic imaginations, has framed a system wherein the sun is supposed to move about the nebula in Orion. But setting aside the extravagant idea of making this luminous spot a center of motion, it must certainly be admitted that the solar motion itself is at least a very possible event."

170. See Letter XVII, p. [243].

171. See the introductory part of his "Account of the Changes that have happened during the last Twenty-five Years in the relative Situation of Double-stars," *Scientific Papers,* vol. II, p. 251.

172. In papers read in 1789 and 1791; for quotations, see my *The Milky Way,* pp. 239 and 241.

173. In his paper "On the Nature and Construction of the Sun and fixed Stars" (1795) Herschel felt himself "authorized upon astronomical principles to propose the sun as an inhabitable world." In the same paper he spoke of the inhabitants of the moon, and praised "the extensive field for animation" which "opens itself to our view" (*Scientific Papers,* vol. I, pp. 479 and 483). In 1798 he referred to the denizens of the satellites of Uranus (*ibid,* vol. II, p. 20), and in 1801 he emphatically re-endorsed his previous statement about the sun by describing it "as a most magnificent habitable globe" (*ibid.,* vol. II, p. 147).

174. As amply evident from the list of his papers and from his extensive correspondence with G. F. Brander, the precision instrument maker in Augsburg.

175. See the opening section of his second memoir "On the Construction of the Heavens" (*Scientific Papers,* vol. I, p. 223).

176. *Bibliographie astronomique avec l'histoire de l'astronomie depuis 1781 jusqu' à 1802* (Paris: De l'Imprimerie de la République. An XI=1803), p. 475.

177. Delambre, *Histoire de l'astronomie au dix-huitième siècle,* publiée par Mathieu (Paris: Bachelier, 1827), p. ix. Among the fifty-six astronomers were Bory, Mason,

Frisi, Marinoni, Pezenas, Godin, Kegler, Bliss, and Long, all certainly deserving of much less attention than Lambert.

178. Paris: Imprimerie du Circle-Social, An IV [1796], vol. II, p. 305. It should, however, be noted that whereas Lambert's "regents" are intrinsically dark, Laplace's "opaque" stars are intrisically luminous, like ordinary stars. Their opaqueness is due to the fact that because of their vastness they had a gravitational field which trapped the light rays emanating from them. Laplace gave the size of such "opaque" stars as 250 times larger in diameter than the sun and having a density equal to that of the earth. Three years later Laplace gave in the *Allgemeine geographische Ephemeriden* a detailed quantitative analysis of the trapping of light rays by such a star, in an article which is available in English translation in S. W. Hawking and G. F. R. Ellis, *The Large Scale Structure of Space-Time* (Cambridge: University Press, 1973), pp. 365–68.

179. See note 159 above. Olbers rejected Lambert's claim about a teleological arrangement of cometary orbits in 1810 in a long article on the possibility of a collision of the earth with a comet. See *Wilhelm Olbers: Sein Leben und seine Werke,* edited by C. Schilling, Volume I. *Gesammelte Werke* (Berlin: Julius Springer, 1894), p. 94.

180. *Populäre Astronomie. Dritter Theil. Physische Astronomie* (St. Petersburg: Kayserl. Akademie der Wissenschaften, 1810), pp. 72–74. Earlier, Schubert stated in his *Theoretische Astronomie* (Zweiter Theil; St. Petersburg: Kayserl. Akademie der Wissenschaften, 1798, p. 57) that "Kepler had already entertained about the plane and shape of the Milky Way ideas similar to those of Lambert and Herschel." In the French translation, *Traité d'astronomie théorique* (St. Pétersbourg: De l'Imprimerie de l'Académie Imp. des Sciences, 1822) even that brief reference to Lambert was omitted. See vol. II, p. 81.

181. *Etudes d'astronomie stellaire sur la voie lactée et sur la distance des étoiles fixes* (St. Pétersbourg: Imprimerie de l'Académie Impériale des Sciences, 1847), pp. 12–16.

182. *Ibid.,* pp. 12 and 16.

183. *Ibid.,* p. 5 of Notes.

184. *Ibid.,* p. 12.

185. F. Arago, *Astronomie populaire,* published by J. A. Barral (Paris: Gide, 1857–58), vol. II, pp. 356–67. For quotation, see p. 357. Arago treated another idea of Lambert, the dark regents of systems of stars, with the same studied indifference to its Lambertian context and specifics. In a half-page-long chapter on "the center around which the stars revolve" Arago wrote that already Lambert had admitted that "stars had general circular movements in immensely large orbits around unknown centers." This movement was in Lambert's eyes, Arago added, "the sole means to give the system of stars perfect stability." (*ibid.,* p. 24). In his massive monograph on comets, *The World of Comets* (translated and edited by J. Glaisher [London: Sampson Low, 1877]), Amédée Guillemin reported Arago's estimate of the number of comets, and again mentioned Lambert and the *Cosmological Letters* as he discussed the question whether comets were habitable (pp. 164–65 and 518–19). Like Arago, Guillemin, too, ignored the Lambertian cosmological background of these questions. According to Guillemin, the *Cosmological Letters* was published in 1765. Arago gave the year correctly, but according to him the place of publication was Leipzig.

186. J. H. Mädler, *Die Centralsonne* (2d revised and enlarged ed.; Mitau and Leipzig: Verlag von G. A. Reyher, 1847), pp. 2–3. The first edition was also published in 1847. Mädler singled out the Pleiades as the central stellar group in the Milky Way,

and located its center of gravity in the vicinity of Alcyon (p. 69). Equally telling was Mädler's silence on Lambert in his *Der Wunderbau des Weltalls oder Populäre Astronomie* (6th ed.; Berlin: Karl Heymann's Verlag, 1867), in which he stressed the teleological aspect of the question about the center of the stellar system (pp. 417 and 439–40). The first edition was published in 1841 in Berlin by the same publisher under the title, *Populäre Astronomie*. (On dark bodies see pp. 376–77, on central bodies see p. 391). Absence of any reference to Lambert was all the more strange because Mädler recalled a great number of astronomers, among them Mayer and Bode, with whom Lambert had close connection. In the detailed survey of the history of astronomy in the sixth edition, omission of Lambert was again conspicuous (see pp. 674–77).

187.   "Die Zukunft der Astronomie" (1840), in *Reden und Abhandlungen über Gegenstände der Himmelskunde* (Berlin: Verlag von Robert Oppenheim, 1870), p. 14. Mädler was already at the end of his career when in his history of astronomy, *Geschichte der Himmelskunde von der ältesten bis auf die neueste Zeit* (Braunschweig: Georg Westermann, 1873) he spoke at some length of Lambert and of his cosmology (vol. II, pp. 216–20). His general evaluation of it was that although subsequent observations had discredited many of Lambert's conclusions, many of his ideas had been completely verified. The only example given by Mädler in this connection was the motion of the sun as the result of its own motion and of the motion of the stellar group to which it belonged (p. 220). Mädler claimed that had Lambert been familiar with the reality of double stars, he would not have insisted on the necessity of central bodies (p. 219). A strange claim, because Lambert emphasized that central bodies were necessary for the stability of triple, quadruple, and multiple stars, but not for double stars! Such inaccuracy on Mädler's part was matched by his statement that Lambert "later expanded and revised the *Cosmologische Briefe* as the *Système du Monde*. Merian published it after Lambert's death in 1779" (p. 217).

188.   *Kosmos. Entwurf einer physischen Weltbeschreibung* (Stuttgart and Tubingen: J. G. Cotta'scher Verlag, 1845–58), vol. I, p. 90. Lambert was again referred to but generically in Humboldt's lengthy discussion of the distribution of stars (vol. III, pp. 143–214; see p. 187). Humboldt's negative attitude toward Lambert is lurking behind his emphatic assertion that the sun is the only central body of which we have positive knowledge (vol. I, p. 93).

189.   See my *The Milky Way*, pp. 258–62.

190.   It is with references to such phrases that begins the account in question, A. Döring's "Kant, Lambert und die Laplacesche Theorie," published in the widely circulating *Preussische Jahrbuch*, 58 (1886), pp. 128–49. Döring dealt separately with Kant's theory of the Milky Way and with his theory of planetary evolution, and he did the same with Lambert's explanation of the Milky Way and with his manner of deriving the structure of the universe. He put in critical light Kant's and Lambert's assertions about the identity of their views, and finally he discussed Laplace's nebular hypothesis as given in the first three editions of the *Exposition*. Döring was fully aware of Lambert's advocacy of a finite universe and of his aversion to an evolutionary outlook. The brief books by R. Zimmermann, *Lambert der Vorgänger Kants: Ein Beitrag zur Vorgeschichte der Kritik der reinen Vernunft* (Vienna: Karl Gerold's Sohn, 1879; see especially pp. 11 and 67) and by J. Lepsius, *J. H. Lambert: Eine Darstellung seiner kosmologischen und philosophischen Leistungen* (Munich: A. Ackermann, 1881; see especially pp. 26–43) are overemphasizing the common traits of Kant's and Lambert's cosmological speculations. It is particularly regrettable in the case of Lepsius that he overlooked the non-

evolutionary character of Lambert's cosmology because his presentation of the contents of the *Cosmologische Briefe* was the best to appear until then and his account of Lambert's life and work was based on serious research. No documentation is given in the address by the famous astrophysicist, Karl Schwarzschild, delivered on Nov. 8, 1907, before the Society of Sciences of Göttingen under the title, "Ueber Lambert's kosmologische Briefe," *Nachrichten von der Königlichen Gesellschaft der Wissenschaften zu Göttingen,* Geschäftliche Mitteilungen aus dem Jahre 1907 (Berlin: Weidmannsche Buchhandlung, 1907), pp. 88–102.

191. For more details, see note 10 to the Preface.

192. The chief example is Lord Kelvin's handling of Olbers' Paradox; see my *The Milky Way,* pp. 276–77.

193. The most appropriate comment about such possibilities seems to be the one which Einstein gave of one such model proposed by himself. It "does not in itself claim to be taken seriously," he remarked in his "Cosmological Considerations on the General Theory of Relativity" (1917). See *The Principle of Relativity* by Einstein and others (London: Methuen, 1923, p. 179).

194. In his book-length discussion of the first volume of Humboldt's *Kosmos* in *The Edinburgh Review,* 87 (1848), pp. 170–229; see especially pp. 184–85. For further discussion of this point in the astronomical literature of the nineteenth century, see my *Paradox of Olbers' Paradox,* chapter VIII.

195. At present the chief promoter of the idea of a hierarchical universe is Gerard de Vaucouleurs. See especially his article, "The Case for a Hierarchical Cosmology," *Science,* 167 (1970), pp. 1203–13.

196. "Wie eine unendliche Welt aufgebaut sein kann," *Arkiv för Matematik, Astronomi och Fysik* (Lund), Band 4. (1908) No. 24. p. 3. Charlier referred to S. Newcomb's *The Stars: A Study of the Universe* (London: John Murray, 1902), in which Newcomb called attention to Lambert's hierarchically ordered universe as a possible solution of its assumed infinity. But Newcomb also noted that "modern developments show that there is no scientific basis for this conception, attractive though it be by its grandeur" (p. 233). In 1922 Charlier offered an even more detailed, mathematical model of a hierarchically ordered system of nebulae, but this time there was no reference to Lambert: "How an Infinite World may be built up," *Arkiv,* Band 16. No. 22.

197. See W. Busch, *Die deutsche Fachsprache der Mathematik: ihre Entwicklung und ihre wichtigste Erscheinungen mit besonderer Rücksicht auf Johann Heinrich Lambert* (Giessen: Giessener Beiträge zur deutschen Philologie, 1933).

198. See Wolf, "Lambert von Mühlhausen," p. 334, on the complaint of contemporaries.

199. Until C. Wolf gave the full translation in French of Kant's *Allgemeine Naturgeschichte* in his *Les hypothèses cosmogoniques* (Paris: Gauthier-Villars, 1886).

200. As ably discussed by R. Jaquel in his "Les conceptions philosophiques et cosmologiques de Lambert." See Selected Bibliography.

201. See Chapter VII, "The Silent Breakthrough," in my *The Milky Way.*

202. The principle of plenitude is the topic of Arthur O. Lovejoy's classic monograph, *The Great Chain of Being: A Study of the History of an Idea* (1936; Harper Torchbook, 1960), where the *Cosmologische Briefe* is discussed on pp. 139–40. On Lovejoy's general conclusion about the intrinsic merit of that principle, see pp. 329–333, where he gives the nod to the principle of limitation or contingency.

II NOTES TO THE *COSMOLOGICAL LETTERS*

PREFACE

1. Lambert's referring as *Photometrie* to the *Photometria,* of which a German translation was published only in 1892, is a small but telling evidence of his intent to be a German purist in his style. See note 177 to the Introduction.

2. The section in question is §1139, where Lambert rejects Chéseaux's model of a homogeneous distribution of stars and states with a pointed reference to the Milky Way that the distribution as observed amounts to the confining of stars within a flattened sphere of which the Milky Way is the ecliptic.

3. Such is an accurate summary of the preponderance of studies connected with the realm of comets and planets over that of the stars. One aspect of this was the almost complete neglect of the Milky Way during the half a century following the publication of Newton's *Principia.* See chapter V, "Newtonian Distraction," in my *The Milky Way.*

4. In the *Photometria* (§1137) Lambert refers with special praise to Huygens' estimate of the distance of Sirius as being 27,664 times greater than the earth-sun distance, but also mentions that on the basis of his measurement of the aberration of light Bradley was entitled to set the minimum distance of stars at 400,000 times that distance. Here Utenhove refers (pp. 36–37) to the memoirs of Herschel on the construction of the heavens, to his second catalogue of a thousand nebulae, and to Kant's speculations in his *Allgemeine Naturgeschichte.*

5. As Darquier notes (p. 38), these rules were published in Lambert's *Logische und philosophische Abhandlungen* (see note 6 to the Introduction).

6. Leibniz merely provded a variation of the principle formulated by Fermat in the 1630's with regard to the refraction of light. For an endorsement by Leibniz in such a context of the principle of teleology, see his "Meditationes de cognitione, veritate et ideis," first published in the *Acta eruditorum* in 1694 *(Die philosophischen Schriften von Gottfried Wilhelm Leibniz,* edited by C. J. Gerhardt, vol. IV [Hildesheim: Georg Olms, 1960], pp. 447–48).

7. Lambert's reference is to Maupertuis' formulation of the law of least action which Maupertuis first enunciated on April 15, 1744 in a memoir read to the French Academy in which he concluded: "We cannot doubt that all things are regulated by a Supreme Being who, while He has imprinted on matter forces which show his power, has destined it to execute effects which mark his wisdom." In his *Essai de cosmologie* (1750) Maupertuis endorsed such general laws as the only reliable pointers to the existence of God. See Philip E. B. Jourdain, *The Principle of Least Action* (Chicago: The Open Court Publishing Company, 1913), pp. 6 and 11. In connection with Lambert's reference to Leibniz and Maupertuis, Utenhove's sole remark is that Laplace took issue with teleology in his *Exposition* (Liv. iii., chap. 2).

8. The remark is aimed at the arbitrary use of teleology for which the followers of Wolff had been increasingly criticized for some time. But as Lambert's next remark shows, he is not pleased with excessive criticism of teleology.

9. The concern in question was heightened by the destruction of Lisbon by earthquake in 1755, an event which dealt a death blow to Leibnizian optimism and to a teleology fostered by it. Yet, doubts about teleology had been on the increase while Leibniz was still alive. They were made the basis of *The Analogy of Religion* (1736) by J. Butler, who defended the thorny aspects of Biblical revelation on the basis that nature, too, failed to appear wholly rational in every respect.

10. What Lambert asserts here is the self-contradictoriness of the idea of an actually realized infinite number, and should be seen in the light of his emphatic assertion in Letter XII of the finiteness of the physical universe. On the former point he echoes an assertion made repeatedly by Leibniz, who, however, seemed to endorse the infinite extension of matter, though not without some perplexity as can be seen in his letters of Feb. 14 and March 17, 1706 to Father Des Bosses (*Die philosophischen Schriften,* vol. II, pp. 301 and 304–05). Lambert's categorical endorsement of the finitenss of the universe is invariably overlooked in histories of cosmology and astronomy, largely under the mesmerizing influence of the "canonization" by Kant of the spatial, Euclidean infinite.

11. Standard references, one would say today.

12. The doctrine of antipodes is the view that the southern half of the earth's surface is uninhabitable, for humans cannot endure a permanent upside-down position.

13. In his *Quaestiones naturales,* Seneca advanced the daring view that comets were not atmospheric phenomena but wandering stars, like the planets, though with orbits which could greatly deviate from the ecliptic. He spoke of the eventual return of each comet, and ascribed our failure to observe any such return to the faintness of their light, or to the fact that we see only the "extremities" of their orbits (Lib. vii, cap. 13 and 23). See the English translation under the title, *Physical Science in the Time of Nero,* by John Clarke (London: Macmillan, 1910), pp. 286 and 296–98.

14. At new moon, the moon is between the sun and the earth, and moves in a direction opposite to the earth's orbital motion.

15. In his *Meterorologica* (Bk. I, chap. iv–vii), Aristotle argues that comets consist of dry exhalations made incandescent below the moon's orbit, the natural place of fire. Since, according to Aristotle, dry exhalations emerge from the earth, and earthquakes can naturally be viewed as triggering the release in large quantities of that exhalation, he postulated an invariable connection between earthquakes and the appearance of comets. It was Tycho Brahe who had first shown conclusively that some comets were superlunary phenomena. His conclusions were heavily exploited by Galileo in his attacks on the Aristotelian world view in the *Dialogues concerning the Two Chief World Systems.*

16. Halley published his Table in 1705. See note III-8.

17. Conversational accounts might have very well exaggerated the number of cases firmly established which the first printed report (see next note) put at fifteen.

18. The "complete induction" was made by Lambert not by Mayer. In the first printed account of Mayer's lecture in the January 21, 1760, issue of the *Göttingische Anzeigen von gelehrten Sachen* (pp. 73–75) mention was made, in addition to the "fifteen," of "many others" whose motion could be presumed. Neither did Mayer attribute motion to all stars, nor did he suggest a systematic motion for them. The latter point is especially to be noted because for Lambert's purposes the systematic motion of stars was indispensable. So was the sun's motion about which the report stated that Mayer's failed to establish anything positive about it. It should seem strange that this account in the *Anzeigen,* although easily available in Augsburg, was never mentioned by Lambert. Mayer's lecture was printed in 1775, and in it the number of stars with certainly established motion was given as 15 to 20. There it was also stated by Mayer that the sun's motion should evidence itself in the separation of stars from one another at a given point in the sky. Mayer's lecture is important because its data provided support for Herschel to reach the conclusion that the sun did indeed move, and in the direction of λ Herculis. The historical background and

importance of Mayer's lecture is concisely set forth by E. G. Forbes in *Tobias Mayer's Opera inedita. The First Translation of the Lichtenberg edition of 1775* (London: Macmillan, 1971), pp. 44–45 and 156–57. For Mayer's lecture, "Treatise on the Proper Motion of the Stars," see pp. 109–112.

19. The typical form of the expression is *reductio ad absurdum* and in both cases it means a chain of reasoning that leads to patently absurd conclusion.

20. A somewhat odd though often repeated expression throughout the work to indicate the need for observational verification of a given proposition.

21. This third main proposition is about the postulated existence of massive dark bodies in the center of star systems.

22. With the exception of Aristarchus, they are listed by Copernicus in his *De revolutionibus* as early proponents of the heliocentric system.

23. King of Castile who around 1250 gathered some fifty astronomers to work out an improved form of the tables of Ptolemy.

24. Lambert refers to the often quoted reflections which Seneca added in defense of his daring thoughts on comets: "The day will yet come when the progress of research through long ages will reveal to sight the mysteries of nature that are now concealed. The day will yet come when posterity will be amazed that we remained ignorant of things that will to them seem so plain." (*Physical Science in the Time of Nero*, p. 298).

25. See Lambert's letter of Feb. 17, 1765, to Frederick the Great quoted in the Introduction and note 97 there.

26. An allusion to Fontenelle's heavy reliance on Cartesian vortices and cosmology.

27. They were at times too lively. In translating Fontenelle's *Entretiens* into German in 1780, Bode remarked in connection with the enormous vivacity of the denizens of Mercury due, according to Fontenelle, to the very high temperature there: "Peculiar! One rather finds with us here [Berlin] that too much heat makes the mind more sleepy and lazy than lively." B. E. Fontenelle, *Dialogen über die Mehrheit der Welten*, mit Anmerkungen und Kupfertafeln von J. E. Bode (Berlin: bey Christian Friedrich Himburg, 1780), p. 190.

LETTER I

1. The extraordinary star is Halley's comet. It began to be called as such from 1759 on when the prediction which Halley had made in 1705 about the return of the comet of 1682 in every 76 years was fulfilled for the first time.

2. Lambert most likely had in mind a passage in Huygens' *Cosmotheoros* (see note X–1) where (p. 140) reference is made to Hesiod's fancy that the earth was from both heaven and hell at a distance which a freely falling anvil traversed in ten days. Such a distance is certainly dwarfed when compared with the distance of a fiftieth magnitude star which Darquier found for obvious reasons so unrealistic that he changed it to the fifth magnitude though adding in a note: "50th in the original, but had this [not] been a printing error, the consequences would be the same" (p. 47).

3. This is approximately the age of creation according to biblical chronology which was still taken by many in Lambert's time in a literal sense. Utenhove notes that it takes 25,000 years for the light to reach us from the most distant nebulae observed by Herschel (p. 48).

4. With the regularity of the comets' paths having been demonstrated in Newton's *Principia*, a most effective scientific blow was dealt to the mythical significance of comets. In the ode which Halley wrote in honor of Newton and which he prefixed to the *Principia*, it is pointedly remarked: "Now we know the sharply veering ways of comets, once a source of dread, nor longer do we quail beneath appearances of bearded stars" (see edition by Cajori, p. xiv). A massive and scathing attack on superstitious belief in comets was made five years prior to the publication of the *Principia* in an almost 600-page long *Lettre* of Pierre Bayle to "M.L.A.D.C., docteur de Sorbonne," in which, as the rest of the title states, "it is proved by several reasons drawn from philosophy and theology that comets are in no way advance signs of any misfortune" (Cologne: chez P. Marteau, 1682).

5. William Whiston (1667–1752) explained the deluge in his *A New Theory of the Earth* (London: printed by R. Roberts for Benj. Tooke, 1696) by the enormous tidal wave supposedly triggered by a previous passing of the comet of 1682 very close to the earth.

6. Thomas Burnet (c. 1635–1715), author of *Telluris theoria sacra* (1680–81) which was published shortly afterwards in English translation as *Sacred Theory of the Earth* (1684–89), painted in vivid colors the great changes which the surface of the earth had undergone since its original formation six thousand years earlier. He assumed no collision of celestial bodies to account for those changes which, however, were catastrophic enough to put Burnet in a class with Whiston and others.

7. Among the "others" the most prominent were Maupertuis with his "Lettre sur la Comète qui paroissoit en 1742" (see *Oeuvres de Mr. de Maupertuis* [new ed.; Lyons: chez Jean-Marie Bruyset, 1756], pp. 209–56); Buffon with his theory of the evolution of the solar system through the grazing collision of a comet with the sun, set forth in the first volume of his famed *Histoire naturelle* in 1749; and not least, Halley who pointedly noted that had the comet of 1680 "come towards the Sun thirty-one days later than it did, it had scarce left our Globe one Semidiameter of the Sun towards the North. And without doubt ... it would have produced some change in the situation and species of the *Earth's* Orbit and in the length of the year . . . [and would have] left this most beautiful order of things be entirely destroyed and reduced into its antient chaos." *Astronomical tables with precepts, both in English and Latin, for computing the places of the sun, moon, planets, and comets* (London: W. Innys, 1752), p. Tttt 4. The passage is from the enlarged English version made by Halley of his "Astronomiae cometicae synopsis" (see note III-8).

8. J. J. Bodmer. See note 79 to the Introduction.

9. The extremely superstitious attitude of the Chinese toward unusual celestial phenomena, such as eclipses and comets, and their meticulous observations of the sky for political prognostication had been well known in the eighteenth century through lengthy reports of Jesuit missionaries.

10. This remark of Lambert is one of the many instances of the general awareness of the "unduly" large distance between Mars and Jupiter, an awareness which led to the formulation of the Titius-Bode law of planetary distances and to the search for a planet between Mars and Jupiter. See note 134 to the Introduction. Utenhove quotes (p. 51) the law without referring either to Titius or to Bode, and recalls that Kant in his *Allgemeine Naturgeschichte* attributed the absence of a planet between Mars and Jupiter to the great size of the latter, and that Kant also traced to this factor the absence of moons around Mars. The two moons of Mars were discovered only in 1877.

LETTER II

1. The intended meaning of the phrase is that one should not think of such possibilities as to dispair of a teleological, providential arrangement of the universe.

2. The ten satellites known in 1761 were the earth's moon, four moon's of Jupiter discovered by Galileo in 1610, and five moons of Saturn, of which Huygens found one in 1655, Cassini two in 1672 and another two in 1684. Utenhove remarks (p. 57) that now [1801] the count is eighteen, including two more for Saturn and six for Uranus.

3. Utenhove remarks (p. 58) that "the satellites of Uranus make an exception. According to Herschel's observations their orbits are almost perpendicular to that of their principal planet." What makes this remark of Utenhove noteworthy is that although he often refers to Laplace's nebular hypothesis set forth in his *Exposition,* he fails to note Laplace's strange silence on this point potentially so destructive of his hypothesis of the origin of the system of planets. See on this my article, "The Five Forms of Laplace's Cosmogony," *American Journal of Physics,* 44 (1976), pp. 4–11.

4. Lambert's judgment was fairly correct. Without taking issue with Lambert, Utenhove refers (pp. 58–59) to data provided by Maraldi, Whiston, and Cassini, and to the recent observations by Herschel and Schröter. To this Utenhove adds his report on some very recent efforts to find a regularity in the relative distances of the satellites of Jupiter similar to the one displayed by the relative distances of planets.

5. As long as Saturn was the most distant of known planets, only Saturn and Mars formed exception to this rule. Taking into account Uranus, Neptune, and Pluto the size is found to decrease with distance beyond Jupiter.

6. As Utenhove remarks (p. 59), Herschel and Schröter found a similar situation for several of the satellites of Jupiter. A century later G. Darwin showed in connection with the moon the tides and the conservation of angular momentum as the factors producing the identity of rotational and orbital periods. But even if one does not have to assume that a comet (the future moon) had to have in advance a rotational period of 28 days, the measure of improbability which Lambert had in mind remains equally enormous.

7. This passage, rather cumbersome in the original, is simply omitted in Darquier's translation, a circumstance which Utenhove does not point out.

LETTER III

1. Halley's comet. Since then a great number of comets have been found with periods shorter than 76 years.

2. Once the speed at aphelion is given together with its distance from the center of the sun, which is one of the foci of the assumed elliptical orbit, the position of the perihelion point can be easily calculated. In this case it would fall within the sun's surface.

3. The book which set a pattern for considerations of this kind was the *Cosmologia sacra* (1701) of Nehemiah Grew, physician and naturalist, who also stressed teleology in his description of the anatomical specimens kept in the Museum of the Royal Society at Gresham College (1681).

4. Lambert's specific reference to insects suggests that he was familiar with F. C. Lesser's *Insectotheologie* (1738) which went through several editions and was translated into

219

French and Italian. Lesser was also the author of *Lithotheologie* (1735), or a discourse on the existence and attributes of God based on the study of rocks. While it may be tempting to smile at the often naive reasoning in such books, it is well to recall that J. H. Fabre, the famed French entomologist, found much support for belief in God from his life-long study of insects.

5. Bernard Nieuwentijdt (1654–1718) was a Dutch physicist, philosopher, and public official. His principal publication, *Het regt Gebruik der Werelt Beschouwingen* (1717), immediately appeared in English translation under the title, *The Religious Philosopher, or the Right Use of contemplating the works of the Creator* (London: J. Senex and W. Taylor, 1718–19), which by 1745 went through its fifth edition. The French translation appeared in 1725 under the title, *L'existence de Dieu démontrée par les merveilles de la nature,* and was carefully studied by Voltaire and admired by Rousseau. In view of Lambert's fondness for Nieuwentijdt's work, it is quite possible that Lambert's fascination with the great nebula in Orion originated in his reading a long passage quoted by Nieuwentijdt (see English translation, vol. III, pp. 813–14) on it from the *Systema Saturnium* of Huygens, the first to spot it in 1656 (see note 41 to the Introduction and note XIII-7).

6. William Derham (1657–1735), Anglican divine and Fellow of the Royal Society, was the author of *Physico-theology* (1713) and *Astrotheology* (1714).

7. This is the dispute between Cartesians and Newtonians. The former relied on vortices which were shown by Newton to be an inadequate mechanism to account for the properties of cometary orbits. Whatever the success of the mathematical formalism offered by Newton, it left unanswered questions about the physical nature of gravitational attraction. Newton himself roundly rejected in the Queries to the *Opticks* and in his letters to Bentley the idea of an action-at-a-distance and endorsed the notion of an intervening medium for the propagation of gravitational influence from one body to another. The Cartesian critics of the *Principia* insisted that it was mathematics and not physics.

8. Halley's Table of comets appeared on p. 1886 of his "Astronomiae cometicae synopsis," *Philosophical Transactions,* 21 (1705), pp. 1882–1904 (See note I-7 for its enlarged English version). For a discussion of its background, contents, and significance, see A. Armitage, *Edmond Halley* (London: Nelson, 1966), pp. 161–67. Halley's Table was reprinted at the end of the *Cosmologische Briefe* without any change, although Utenhove seems to imply the contrary (p. 66). Utenhove then states that the best up-to-date catalogue of comets is the one added by Zach to Olbers' treatise of computing the orbit of comets (see note 159 to the Introduction). It is to this catalogue of Zach that many of Utenhove's subsequent references are made.

9. The one occurring three times is Halley's comet (1682, 1607, 1531), and Halley himself made that identification in his "Synopsis" (p. 1897). For the one occurring twice Lambert probably had in mind the entries for 1532 and 1661, but as Darquier remarked (p. 68), such an inference was uncertain on the basis of Halley's data.

10. Utenhove gives (p. 67) the corresponding data from Zach's catalogue.

11. Utenhove notes (p. 68) that of the 20 comets which according to Zach's catalogue have their perihelia within the orbit of Mercury there are three with inclination less than 30 degrees.

12. The one in question is Newton. See *Principia,* Part III, Prop. XLII, Problem XXII (p. 540 in Cajori's edition).

13. Utenhove is quick to add (p. 68) that it can be said with certainty of only one comet that it had been observed repeatedly: "The identity of several others conjectured by various astronomers is today rather doubtful."

14. The comet of 1680, one of the most spectacular ever to appear, approached the earth, Utenhove reports (p. 68), to within 20 times the earth-moon distance.

15. In referring to Uranus, which is smaller than Saturn, which in turn is smaller than Jupiter, Utenhove notes (p. 69) that most likely there is a maximum for the mass of planets and that accordingly the mass of comets, too, might decrease beyond a given distance. One has, Utenhove adds, to compare with this Kant's views in his *Allgemeine Naturgeschichte* (Part II, chap. 2).

LETTER IV

1. An erroneous view as was already pointed out by Utenhove. See note III-14.

2. Lambert should have perhaps written "slightly different" in order not to weaken the argument in support of the return of at least two comets on the basis of the same Table.

3. Utenhove reports (p. 72) the calculations of Dionis du Séjour, who showed in his *Essai sur les comètes* (pp. 207–08; see note 123 to the Introduction) that in some specific cases a comet can become the satellite of a planet for a time, but Utenhove insists on the extreme improbability of such an event.

4. The reference is to Halley's comet which stays beyond Saturn's orbit for much of its 76-year-long period.

5. In view of the six tails developed by this comet the example was not only clear but also spectacular. Utenhove notes (p. 74) that its period was calculated by Euler as 122,683 years, whereas Pingré found it to be only 21,808 years. "From this one can see the smallness of certainty that we have even today, and we shall certainly have for a long time, on the periodic revolutions of these comets," Utenhove adds with a touch of resignation (p. 74).

6. As Darquier remarks, these figures should be taken in reference to the earth-sun distance as being equal to 100,000. Utenhove in turn notes (p. 75) in a somewhat benevolent reasoning that the updated catalogue of comets confirms Lambert's opinion.

7. Here (p. 76) Utenhove gives a tabulation of the updated list of 94 comets and finds a complete agreement with Lambert's analysis of Halley's Table.

8. See note 4 to the Preface. The ratio of the Saturn-sun distance to the earth-sun distance is slightly less than 10.

9. Utenhove remarks (p. 77) that on the basis of the updated catalogue the corresponding number would increase to 12,000.

10. According to Utenhove (p. 77) this is the assumption which underlies Lambert's conviction about the increase of the number of comets with the square of distance.

11. Among such aerial phenomena are halos and meteors. Their list as given in Book I of Aristotle's *Meteorologica* was generally accepted until the middle of the seventeenth century which saw the publication of the last commentaries on the *Meteorologica*. It is precisely this circumstance which renders unreliable the older parts of chronological listings of comets. Among such lists Utenhove mentions (p. 78) the ones compiled by Lubeniezki, Riccioli, Hevelius, and Pingré.

12. See note I-9.

13. To allay the shock of this apparently enormous number, Utenhove notes (p. 78) that on the basis of considerations more restricted than those of Lambert "64,000 million" comets were projected by J. F. Wurm in 1787 because of the much greater number of comets with calculated orbits. According to Wurm the number projected on the basis of Lambert's assumptions would be $5\frac{1}{2}$ billion [million million]. In referring in his article, "Verschiedene astronomische Bemerkungen . . ." (*Astronomisches Jahrbuch für 1790* [Berlin, 1787, pp. 169-73]), to Lambert's method of estimating the number of comets, Wurm did not mention Lambert's cosmological ideas (see p. 169), a detail not mentioned by Utenhove. See also note IX-5.

LETTER V

1. The claim that God had to create the best of all possible worlds was most memorably endorsed by Leibniz, especially in his *Theodicy* and in his correspondence with Clarke.

2. The "many of our philosophers" are the followers of Whiston who spoke of comets as "unform'd planets" in his *New Theory of the Earth,* p. 75 (see note I-5).

3. These data and comparisons reflect both the vagueness prevailing in these matters in Lambert's time and also his interest in the physics of heat of which the major evidence is his *Pyrometrie,* or a treatise "about the measure of fire and heat," published posthumously in 1779.

4. Whiston's principal concern was the explanation of the deluge. He offered but scattered remarks of more basic and more general cosmological portent. He seemed to hold that originally God created fiery and opaque matter and that both were formed by God into separate lumps, the stars and comets. The latter were also put by the Creator into very elongated orbits around the former. From there on Whiston's explanations were purely mechanical. Through near-collisions some of the still chaotic comets obtained fairly circular orbits around the stars, and due to a steady exposure to their heat they became planets. Further adjustments of planetary orbits, of periods of rotation and the like were also to be effected through near collisions with comets. For further details, see my forthcoming book, *Planets and Planetarians: A History of Theories of the Origin of Planetary Systems,* chap. IV, "Convenient Collisions."

5. The title of a celebrated work of Ovid heavily influenced in its cosmogonic part by Hesiod's *Theogony* in which the primordial changes of the world are described in terms of feuds among members of the Olympian family of gods.

6. See notes 15 to the Preface and IV-11.

7. Kepler and Hevelius followed Tycho Brahe in viewing comets as condensations of the superlunary ether. See note 15 to the Preface.

8. Utenhove remarks (p. 84) that the total count is 25 which includes Uranus and its satellites. See note II-2.

9. A description which echoes the Biblical and geocentric view of the world.

10. See note IV-10.

11. The choice of 12 may have to do with the fact that one can place around a sphere only 12 similar spheres. Darquier translates Lambert's phrase as a hypothetical proposition (p. 86).

12. A most forceful exponent of this view was Leibniz for whom even every monad was teeming with life. See his *Monadology, §§66–69*.

13. A paraphrasis of expressions often occuring in the Bible.

14. The ability in question was called in doubt by Merian in his condensation (1770, p. 82), and he was echoed by Utenhove (p. 88). In view of the rapidly changing position of those super-astronomers living on comets, both Merian and Utenhove doubted that they could observe even as much as astronomers can on the earth.

LETTER VI

1. Utenhove refers (p. 92) to C. Maclaurin's famed *An Account of Sir Isaac Newton's Philosophical Discoveries,* first published in 1748, in which Maclaurin discussed in general terms and with no reference to any Proposition in Newton's *Principia* the consequences of any deviation from the inverse square law of gravitation. Newton dealt with this question in quantitative details in Book I, Prop. XLV, Problem XXXI (see Cajori's edition, pp. 141–47) and found, as he put it in Book III, Prop. II, that "a very small aberration from the proportion according to the inverse square of the distances would produce a motion of the apsides sensible enough in every single revolution, and in many of them enormously great" (*ibid.*, p. 406).

2. A good example of the licence taken by Darquier is that here (p. 94) he makes Lambert speak of *the* four elements [of Aristotle] and of the need of investigating them further. Utenhove, without seeing the injustice done to the original, adds a long note debunking the Aristotelian restriction of the number of elements to four.

3. The idea that the center of the earth was still in a fiery state was given much credence and publicity through Buffon's cosmogony in the first volume of his *Histoire naturelle.*

4. The computation in question refers to the temperature of the comet of 1680 as estimated in the *Principia* where Newton wrote: "And therefore the heat which dry earth on the comet, while in its perihelion, might have received from the rays of the sun, was about 2000 times greater than the heat of red-hot iron" (see Cajori's edition, p. 521).

5. Needless to say, the absolute cold mentioned by Lambert has nothing to do with the notion of absolute zero introduced by Lord Kelvin a hundred years later. 1761 was the age of caloric and even the measurement of ordinary temperatures was still beset with uncertainties.

6. The "50,000 years" prompted Utenhove to remark: "One may doubt this proposition in view of the ignorance where we still are concerning the laws of the increase and decrease of heat in bodies" (p. 96).

7. Utenhove asks (p. 96) whether the sun is really a hot body, or whether it is merely enveloped in a fiery layer. Underlying this question is the caloric theory of heat which makes Utenhove state that "the rays of the sun can heat the earth only because of their action on the caloric material which is in the earth and in its atmosphere." The absence of enough caloric explains therefore the very low temperature in very high mountains, although directly exposed to the sun, and the cold in deep caverns. "Paradoxical as this assertion may appear, it is subscribed to by great astronomers and physicists," Utenhove adds, and refers to Bode, de Hahn, Schröter, Herschel, de Luc, Erxleben, Lichtenberg, and even to a memoir of Lavoisier and Laplace in the *Mémoires* of the Académie (1780, p. 335). For Lambert the sun was an intrinsically

hot and luminous body existing for the sole purpose of supporting life not only on earth but on all planets and comets.

8. A good instance of the risk involved in declaring in science something specific to be impossible. Uranus is twice as distant as Saturn and twelve times smaller, a remark pointedly made by Utenhove (p. 99).

9. Utenhove refers (p. 99) to Kant's speculation in the *Allgemeine Naturgeschichte* (p. 52), where Kant assumed that the eccentricity of planets increased with distance and that there was a gradual transition from planets to comets in that respect. Utenhove then reports Laplace's speculation in the *Exposition* (vol. II, pp. 301–02), where Laplace explains the great eccentricities of cometary orbits as an evidence of the originally extended solar atmosphere into which all nearby comets with less eccentric orbits were supposedly absorbed.

10. In a long note (pp. 99–100) Utenhove rejects as an "astrological delirium" the ancient Egyptian belief, reported by Diodorus of Sicily, in the return of comets; gives credit to Seneca; states that the prediction of the return of comets can be done either geometrically or empirically; and defends the notion of the return of comets in spite of the failure to observe in 1789 the expected return of the comet of 1661.

11. Halley's comet, whose period is about 76 years. Darquier notes (p. 101) that Lexell had estimated the period of the comet of 1770 as five and a half years.

12. See note 4 to the Preface and note IV-8.

13. Utenhove notes (p. 102) that the orbit of Uranus is less eccentric than that of Jupiter.

14. This is the product of 6 with the ratio of Saturn's distance from the sun to that of Mercury.

15. A perfect case of good prediction for the wrong reason. It must, however, be noted that whereas the distances between planets beyond Saturn still increase, they do not do so, with the exception of the Saturn-Uranus distance, in the same proportion. See note 134 to the Introduction.

LETTER VII

1. Two great imaginary circles, one of which passes in a perpendicular direction through the ecliptic at its points of summer and winter soltice; the other crosses the ecliptic at the two equinoctial points.

2. At this point Utenhove refers (p. 109) to Laplace's evaluation in the *Exposition* of the high degree of improbability that the orbits of all planets and satellites should be in essentially the same plane, if such an arrangement were to be ascribed to mere chance. What makes this remark of Utenhove noteworthy is that he proposes a slight change in the figure given by Laplace in view of the heavy deviation of the orbital planes from the ecliptic of the two satellites of Uranus discovered by Herschel in 1787. Ten years later Herschel announced the discovery of four more satellites of Uranus and stated their retrograde motion, a fact of which Utenhove does not seem to have been aware in 1800 and which Laplace continued to ignore. For further details, see my article, "The Five Forms of Laplace's Cosmogony," quoted in note II-3.

3. Mars, Jupiter, and Saturn.

4. Utenhove notes (p. 111) the almost complete disrepute into which teleology had fallen by then. It is considered, he remarks, "as arbitrary and chimerical as astrology." But, Utenhove continues, some of the "most famous philosophers of our century think

otherwise," and he quotes an article of Kant on the use of teleological principles in philosophy, published in 1788, and the sixth Preliminary Discourse in J. A. de Luc's *Lettres physiques et morales sur les montagnes et sur l'histoire de la terre et de l'univers* (En Suisse: Les Libraires Associés, 1778). De Luc was hardly one of the "most famous philosophers of the century" and the endorsement of teleology by the post-critical Kant in his "Über den Gebrauch teleologischer Principien in der Philosophie" (*Kant's Werke,* vol. VIII [Berlin: G. Reimer, 1912], pp. 159–84) could in ultimate analysis only be a rejection in disguise.

5. As Lambert's own case shows, this method can easily turn into a full circle.

6. In Letter VI.

7. To what extent wishful thinking can influence the reading of scientific and philosophical texts is well illustrated by Utenhove's note (p. 112). Although he reports T. Mayer's firm negative view about the existence of atmosphere on the moon, he makes much of Euler's "suspecting" its existence and of Schröter's observation of a "residual" atmosphere there. Of ideas capable of befogging clear judgment few exerted as pernicious an influence as the notion of denizens on other planets. Schröter himself was a case in point; see Introduction and note 163 there.

8. It is again the uncritical bent of mind, characterizing almost invariably hopeful speculations about denizens on other planets, which transpires in Utenhove's reference (p. 113) to Huygens' *Cosmotheoros* and Fontenelle's *Entretiens* as works in which "reasonings" are given about intelligent beings very different from us. It makes a telling reflection on the "rationalism," "enlightenment," and "critical philosophy" of the eighteenth century that "reasonings" which were sheer fictions were looked upon with a most serious mien.

9. A curious phrase, as the "body" of no comet did ever appear with the same distinct contours as do the planets, and therefore the "body" in question was never observed in the sense intended.

10. A major source of information on that comet was the monograph of Chéseaux who eagerly looked for its phases but found none, as Utenhove remarks (p. 114), in clear indication of the fact that defining the bright nucleus of the comet as its body was largely hypothetical.

11. Here Lambert echoes a view already stated by Newton, who argued in the *Principia* (Book III, Prop. XLI, Problem XXI) that either both the head and tail of the comet fall into the sun or neither does.

12. In this concluding section the French translation degenerates into something even less than a paraphrasis. The corresponding sentence is given by Darquier as follows: "I believe that comets are in general very varied; their volume must be proportional to the extent of their orbit." The singling out of this phrase was motivated by the note added to it by Utenhove (p. 116). He remarked that the prevailing opinion in his day was that comets have a very small mass. The reason for this was the recognition that whereas not even the largest comets were found to affect the orbits of planets, these produced considerable changes in the orbits of the comets of 1759 and 1770. There are even some observations, Utenhove continued, which suggest that some comets have no solid body at all. To this Utenhove added as a concluding remark that "the great majority of comets must have a very considerable volume especially in view of their atmospheres and tails" (p. 116). Such a conclusion was unobjectionable, but also expressive of the wishful thinking which wanted to see comets as being inhabited in spite of all plain indications to the contrary. In fact, already in Newton's time a fair estimate could have been made of the mass of comets and on its

basis dreams about denizens on them might have been laid to rest once and for all. This was not, however, attempted. As late as 1784, Pingré, as already discussed in the Introduction, could blithely overlook the obvious.

LETTER VIII

1. The implication is that they will eventually be so powerful. The theoretical limits of magnification in telescopes were not systematically studied until a century later, and the results put beyond the realm of possibility ever seeing through telescopes living beings, even at the distance of the moon. The sensation created in 1835 by a series of fictitious reports in the *New York Sun* that Sir John Herschel had observed among other things "continuous herds of brown quadrupeds, having all the external characteristics of the bison, but more diminutive than any species of the *bos genus* in our natural history" (see chapter X, "Light on the Moon by the New York Sun," in P. Leighton, *Moon Travellers* [London: Oldbourne, 1963], p. 174) is an ever useful reminder of how misleading hopes of this kind can become.

2. As Darquier noted (p. 118), by the time his translation was being done the number of comets with calculated orbits was about eighty.

3. Utenhove refers (p. 218) once more to the approximately five years which Lexell specified as the orbital period of that comet.

4. Lambert obviously has in mind an optimum value, broader in meaning than the one implied in computations of maxima and minima which attracted much interest since the publication of Maupertuis' formulation of the law of least action. See note 7 to the Preface.

5. As already discussed in Letter VI. See note 1 there. Utenhove, after explaining the consequences of central forces for the motion of bodies in one of the conic sections, suggests (p. 120) that the central force obeying the inverse square law is most natural if the force is to act uniformly in every direction, because the surfaces of concentric spheres increase as the square of distance.

6. The periodic ones are the circle and the ellipse, whereas the parabola and hyperbola are not.

7. As an example Utenhove refers (p. 122) to the comet of 1759 which has a period equal to that of Uranus, but an orbit which carries it twice the distance of that planet from the sun.

8. Darquier gives (p. 122) the etymology of the word as "to see with one's own eyes."

9. Whereas the number of comets actually observed had greatly increased during the generation following Lambert's death and even more so during the last 150 years, and whereas the number of comets with computed orbit was over 800 by 1962 and can only increase rapidly with the wider application of computer techniques, comets have turned out to play a role very different from the one which Lambert expected to be fully verified from the increase of their number through further observations.

10. Here Lambert refers to the fact that elliptic, parabolic, and hyperbolic orbits can at times be very similar to one another since only their sections near their common focal point, the sun, are subject in most cases to observations.

11. The expression, "new creation," need not necessarily mean a creation out-of-nothing but only a major transformation, especially through collisions of celestial bodies. Its rejection by Lambert is a clear indication of his anti-evolutionist outlook.

12. This view fully reflects Newton's conviction as expressed in the Scholium in the *Principia,* and was further elaborated by R. Bentley and other divines as an evidence

226

of the existence of God. The theories of Buffon, Kant, and Laplace on the evolution of the solar system were alike an effort to show that there was no need to resort to a direct action of God to account for the present arrangement of the solar system.

13. Utenhove remarks (p. 125) that this essential sameness of the plane of rotation of the sun and of the planets was a principal evidence for Laplace about their common origin.

14. Such a strange view, which in retrospect may appear almost absurd, was a by-product of the caloric theory of heat. The post-critical Kant defended it spiritedly in an article written in 1794 on the moon's influence on the temperature of the earth's atmosphere: "Etwas über den Einflusz des Mondes auf die Witterung," in *Kant's Werke* vol. VIII, (Berlin: G. Reimer, 1912), pp. 317–24. See also note VI-7.

15. Instead of "die Bewohner der Weltkörper" the original reads "die Bewohner der Weltbürger," an obvious misprint.

16. Contrary to Lambert's claim no observations had been made of the "body" of any comet. See notes VII–9 and VII–12.

LETTER IX

1. In his article, "Observations of the Appearances among the Fix'd Stars, called Nebulous Stars" (*Philosophical Transactions,* 38 [1733–34], pp. 70–74), Derham took the view that several nebulous stars, including the great nebula in Orion, may be "Chasms or Openings into an immense Region of Light, beyond the Fix'd Stars" (p. 74). Derham was ready to identify this "immense Region" with the *coelum empyreum,* endorsed, as he put it, by "the Learned in all Ages." Utenhove recalls (p. 129) the discovery of the great nebula in Orion by Huygens in 1656 and various reports submitted between 1771 and 1796 among others by Messier and Schröter (see Introduction and notes 161–63 there) on the variability of its light and shape. The implicit thrust of Utenhove's note seems to be a confirmation of Lambert's subsequent insistance that the nebula in Orion was the surface of a dark regent illuminated by the surrounding stars, but Utenhove also reports Herschel who, although unable to resolve that nebula into stars, saw in it a part of the Milky Way.

2. "Weltkugel" (celestial globe or globe) stands for planets as well as comets.

3. Comets with hyperbolic orbit.

4. A somewhat inaccurate generalization. Nicholas of Cusa most emphatically assigned denizens to each celestial body, including the sun, in his *De docta ignorantia* (1440), the most frequently printed cosmological book until the late sixteenth century.

5. This number is essentially of the same order of magnitude as the 5 million calculated in Letter IV. See note 13 there. In 1952 J. G. Porter concluded from an analysis of the perihelion distances that there are 5 million comets with their perihelia within Neptune's orbit. Somewhat earlier J. van Oort arrived at the figure of 100 billion (10^{11}) as the number of comets that are around the sun within a distance 5000 times greater than Neptune's distance from the sun. See N. B. Richter: *The Nature of Comets,* translated by A. Beer, with an Introduction by R. A. Lyttleton (London: Methuen, 1963), p. 5.

6. Yet Lambert will confidently speak of the eventual observation of dark regents, far more distant in his estimate than comets with satellites.

7. See note VII-2.

8. Above all, Newton in the *Principia.* For a detailed discussion of the whole history of this question, see E. J. Aiten, *The Vortex Theory of Planetary Motions* (London: Macdonald, 1972).

9. A point already emphasized in the previous Letter where Lambert spoke also of the plane of the sun's rotation as being essentially the same as that of the rotation of planets. Utenhove briefly cites (p. 134) Descartes, Jean Bernoulli, Bouguer, Buffon, Kant, and Laplace as ones who looked for a physical explanation of this identity.

10. According to Utenhove (p. 135), the apparent disorderliness of cometary orbits proved to be an embarrassment to those concerned with cosmology, among them Kant who tried to see a mere visual effect in the retrograde motion of many comets (*Allgemeine Naturgeschichte,* p. 58).

11. See note VIII–4.

12. As Utenhove remarked (p. 137), Kant had already insisted on such a connection of space and time in his *Allgemeine Naturgeschichte* (p. 7).

13. What follows here on the fixed stars and the Milky Way is an artful invitation to Lambert on the part of his fictitious correspondent to give his own explanation of the Milky Way in the next Letter. Here only its visual appearance is described.

14. If space and time are connected, then motion must be attributed to the fixed stars; hence the possibility of some of them colliding with the sun, a collision far more disastrous than that of a comet with the earth by which the "philosophers" mentioned in Letter I had set great store.

LETTER X

1. A fate that had already overtaken Descartes' cosmology. The first to speak of it as nothing more than a beautiful novel was possibly Huygens. See the English translation of his *Cosmotheoros,* published almost simultaneously with the Latin original under the title, *The Celestial Worlds Discover'd* (London: printed for Timothy Childe, 1698), pp. 157–59, and his manuscript note Nr. 2791 in *Oeuvres complètes de Christiaan Huygens* (The Hague: Martinus Nijhoff, 1888–1950), vol. X, p. 403.

2. Hipparchus (fl. 2d century B.C.) was an astronomer from the island of Rhodes. His catalogue comprised 850 stars. He is also credited with the discovery of the precession of the equinoxes, of the eccentricity of the sun's (apparent) orbit, and of some of the inequalities of the moon's motion.

3. A considerable oversight on Lambert's part, as Halley in his effort to re-determine the precession of equinoxes found that the "Latitudes of three of the Principal Stars in Heaven [Aldebaran, Sirius, Arcturus] . . . contradict the supposed greater Obliquity of the Ecliptick." Since he knew that three great observers, Timarchus, Hipparchus, and Ptolemy, were in agreement in fixing the position of these stars, and on the other hand he could not believe that they could be in error of almost half a degree, the conclusion seemed to be inevitable that three stars were in motion. After all, they were the "most conspicuous," and "in all probability nearest to the Earth!" Furthermore, a motion which could not be observed "in the space of a single Century of Years" could very well be verified "in so long a time as 1800 Years" through "the alteration of their places." This was an argument which, in Halley's words, "seems not unworthy of the Royal Society's consideration, to whom I

humbly offer the plain Fact as I find it, and would be glad to have their Opinion." See his "Considerations on the Change of the Latitudes of some of the principal fixt Stars," *Philosophical Transactions,* 30 (1718), pp. 736–38. Lambert's oversight is not pointed out explicitly by Utenhove who in a long note (p. 143) mentions from Halley to Mayer all the major contributors to the elucidation of the motion of stars.

4. Utenhove remarks (p. 145) that Kant had already made this assumption in his *Allgemeine Naturgeschichte* (p. 102).

5. According to Utenhove (p. 145) Kant derived from his system the intrinsic luminosity of such a central body (*ibid.,* pp. 131 and 138).

6. By "general reason" Lambert means a common, underlying mechanism, operative everywhere in nature. In the non-evolutionary view of Lambert one such mechanism could be the vortex to which he will return in Letter XX. In an evolutionary view, such as Laplace's nebular hypothesis, this common mechanism offers itself naturally, though not in a way which is free of great difficulties when one takes into account the distribution of angular momentum and mass between the sun and the planets.

7. Here Utenhove claims (p. 146) that the first to demonstrate this proposition was Lalande in 1776. The question was, as Utenhove reports, discussed by Schubert, Laplace, Klugel, and Wurm. Contrary to the impression created by Utenhove, Lambert's proposition was not proved at that time, or later.

8. This is Lambert's first statement about the distinctness of the visible stars from the Milky Way, a detail which is one of several that distinguish his notion of the Milky Way from that of Wright and Kant.

9. In his translation Darquier speaks (p. 149) of the zodiac, not of the ecliptic. Utenhove recalls (p. 149) Kant's astonishment (*Allgemeine Naturgeschichte,* p. 9) that such an explanation had not occurred to anyone before. Utenhove then mentions Wright and Herschel together with a reference to Kepler as their possible forerunner. The early history of speculations about the Milky Way was of course far richer, as documented in my *The Milky Way,* see especially chapters IV and V.

10. Utenhove aptly notes (p. 149) that Lambert's notion of the Milky Way does indeed differ from that of Kant for whom the sun immediately belonged to the Milky Way.

11. This categorical assertion by Lambert of the resolvability of the whiteness of the Milky Way into stars contrasts sharply with the hesitation of others at that time concerning the true cause of its whitish hue, of that of other nebulous stars, and in particular of the Magellanic Clouds. In his report of his observations made in 1751–52 of the nebulous stars of the southern hemisphere, "Sur les Étoiles nébuleuses du ciel austral," (*Mémoires de l'Académie Royale des Sciences* [Paris: de l'Imprimerie Royale, 1761], pp. 194–99), which he communicated to the Académie in 1755, Lacaille took the view that it was not certain that the most whitish parts of the Milky Way are caused "as commonly believed by a heap of small stars more crowded there than in other parts of the sky" (p. 195). As is well known, a generation later Herschel began a struggle of thirty years to clear up the doubts about the resolvability of nebulae into stars, a struggle that ended inconclusively. While Utenhove asserted (p. 150) on Herschel's authority the full resolvability of the Milky Way into stars, Darquier still insisted (*ibid.*) that Lambert's assertion of the same resolvability was contradicted by observations.

12. Most likely Lambert had in mind the protrusion near Cepheus. See *Almagest,* VIII, 2.

13. An evaluation which cannot be made rigorously. Lambert's estimate is much too large.

LETTER XI

1. A reference to T. Mayer's investigations. See note 18 to the Preface.

2. Utenhove remarks (p. 154) that such was already the procedure of Kant who in turn followed Wright.

3. See note 4 to the Preface and note IV–8.

4. That the stars composing the Milky Way are very small was originally put forward by Democritus. Such was also the majority view during the Middle Ages when Aristotle's explanation of the Milky Way as a sublunary cloud was one of his few tenets that failed to attract a general following. References to the Milky Way as composed of small stars were often present even in the eighteenth-century astronomical literature (see note X-11). Lambert's view has, therefore, a measure of originality.

5. See note 3 above.

6. This notion is particularly emphasized in Wright's account of the Milky Way and is also present in that of Kant. Once more Darquier renders ecliptic with zodiac (p. 156).

7. Utenhove aptly notes (p. 157) that Herschel did not find such a great disproportion. Whereas the ratio as given by Lambert is 1 to 10,000, the one implied in the diagram of the Milky Way given by Herschel in 1785 corresponds to a ratio of about 1 to 5.

8. The allusion is to the complicatedness of Oedipus' story.

9. As Lambert himself put it, this model was merely "imagined" by him. Perhaps with lamps that are not point sources of light the need for a very large spherical shell empty of them could be justified in order to let that "break" appear.

10. The question was not only unexplored, it was not even posed except by Gassendi and Guericke, who for an answer referred to the unfathomable wisdom of the Creator. See my *The Milky Way,* pp. 123 and 134.

LETTER XII

1. In his letter of November 13, 1765, to Kant (see note 38 to the Introduction) Lambert specifies that night as having occurred in the summer of 1749.

2. There are a few references to astronomical matters in the little notebook which Lambert filled mainly with elementary exercises relating to arithmetic, geometry and chemistry, before he went to Basel as a sixteen-year-old boy. The notebook is in the Universitätbibliothek Basel. Georges Rémy gave a wholly misleading account of this notebook by describing it as a "vrai memorandum de toutes les matières possible, classées avec ordre et méthode" in his well-intentioned but often uncritical article, "Jean-Henri Lambert: sa vie et son oeuvre" (*Revue d'Alsace,* 61 [1910] pp. 393–406 and 452–68); for quotation, see p. 396, where Rémy also stated that at the age of fifteen Lambert "observa pour la première fois le trajectoire d'une étoile!" Rémy's four-page summary of the contents of the *Cosmologische Briefe* (pp. 457–61) is marred with similar errors, among them his assigning of its composition to the late 1760's.

3. A clear indication that not all the major features of his theory of the Milky Way occurred to him at once.

4. Lambert's remark reflects the uncertainty which at that time was particularly felt about the reliability of some claims, especially about those made by Horrebow, concerning

5. Thanks to spectroscopy, the velocity of stars began to be measured almost within one century.

6. The original reads "zunimmt," instead of "abnimmt."

7. Utenhove notes (p. 165) that a star as bright as the sun, but at a distance where its parallax would be 1 second, would have an apparent diameter of $\frac{1}{200}$ second. He also notes Herschel's success of measuring the apparent diameter of some stars as being of about 1 second. Then Utenhove refers to Laplace's calculation of the effect of the gravitational field of a star 250 times larger than the sun on its own light. See note 178 to the Introduction.

8. Lambert uses "Himmelsluft," another indication of his intent to avoid even such common foreign words as the ether.

9. See notes VI-7, VIII-14, and 173 to the Introduction.

10. Very approximately indeed, as simple calculation would show. The progression, widely referred to up to that time, was wholly discredited, Utenhove notes (p. 166), still during Lambert's life through the thousands of stars catalogued by Lacaille and Lalande according to magnitude.

11. This is the "close-packing" model of stellar distribution. See note 12 to the Introduction.

12. The reference to geometry shows that Lambert was aware of the fact that 12 points, as Utenhove explains in a long note (pp. 167–68), cannot be placed around a sphere in a way which would be symmetrical in every respect.

13. See note XI-2.

14. See note 4 to the Preface and note IV-8.

15. Computing the time needed for the collision of two stars would be different from computing the earth's fall into the sun (see Letter X) on the basis that the sun remained at rest.

16. In view of Lambert's enthusiastic acceptance of the news about Mayer's feat, the intended meaning of this sentence is most likely the great difficulty of ascertaining that change because of its smallness.

17. The "very different distance" should be understood in reference to the smallest distance estimated to be 500,000 times the earth-sun distance.

18. This idea of systems of dark bodies could not, at any rate, be consonant with purpose.

19. See note X-2.

20. A perfect illustration of lack of clarity in Lambert's style. Instead of using the adjective "wide," he could simply have spoken of "diameter."

21. Lambert's emphatic assertion of the finiteness of the universe. See note 10 to the Preface.

22. See note IX-1.

23. This point will be argued in detail in Letters XVII and XVIII. See also note 41 to the Introduction.

24. A happy conjecture with respect to the peripheral position of our local system of stars in the Milky Way. Herschel put the sun close to its center, a position from which it was not removed until the 1920's.

25. The real cause of these variations, called secular variations, was, as Utenhove noted (p. 172), cleared up by Laplace and Lagrange, shortly after Lambert died. They are due to the mutual perturbation of the planets.

26. This is the first intimation of an idea, the cycloidal character of each orbit, which Lambert will develop in the concluding Letters.

27. As Utenhove remarked (p. 173), within twenty or so years following Lambert's death the margin of error was reduced, largely through the work of Delambre, to $\frac{1}{3}$ minute.

28. This was the principal target of the work of de la Hire, Lacaille, and d'Alembert in the 1730's and 1740's.

29. Here Utenhove remarked (p. 174) that the cause of this should be sought in the mutual perturbation of planets and not in the influence of a dark central body.

LETTER XIII

1. The "primum mobile," or the starry firmament as a spherical shell, was a basic tenet only in the Platonic-Aristotelian cosmology. The atomists and the Epicureans viewed each star as an independent world subject to a perennial cycle of formation and dissolution. Among these there were no notable astronomers, all of whom, either for philosophical or for computational reasons, endorsed the notion of a universe enclosed within its spherical "primum mobile."

2. Such is an oversimplified account of the endorsement by Ptolemy and by his predecessors and followers of a spherical, closed universe, as can be seen from even a cursory look at the introductory chapters of Book I of his *Almagest*. Had the case been as simple as Lambert suggested, Copernicus would not have had to present in Book I of his *De Revolutionibus* a meticulously constructed chain of arguments to discredit Ptolemy's position.

3. With Herschel's large telescopes the relative displacement of some double stars could indeed be ascertained "in a few" years. See his report (note 171 to the Introduction) of 25 years of observations of double stars.

4. The intended meaning here is not so much that stars in the Milky Way are crowded but that all are in motion.

5. In view of what Lambert had already said about Mayer's results, he might have expressed himself more positively.

6. According to Utenhove (p. 177) an inaccurate rendering by D. Gregory in his *Astronomiae physicae et geometricae elementa* (Lib. iii, Sec. ix, Prop. 54) of what G. D. Cassini (1625–1712) wrote of γ Aries as being the visual fusion of two fifth-magnitude stars, gave rise to the widespread notion that Cassini had claimed that he had observed it now as one, now as two stars. See pp. 273–74 in Gregory's work published in Oxford in 1702.

7. Huygens reported this observation of his in his *Systema Saturnium* (1659). See *Oeuvres complètes,* vol. XV, p. 239. With the improvement of telescopes many more stars were spotted there. As Utenhove noted (p. 178), Schröter showed in his *Aphroditographische Fragmente* (p. 48) that their relative displacements cannot be attributed to the earth's motion. Herschel himself touched on this in his paper, "On the Parallax of the Fixed Stars," (1782), *Scientific Papers,* vol. I, p. 48.

8. In other words, with a system in which the motion of stars is around a huge dark central body and not merely around their common center of gravity.

9. Lambert perhaps had in mind "the very exact solar and lunar tables" published a few years earlier in the *Mémoires posthumes de F. Ph. Loys de Chéseaux sur divers sujets d'astronomie et de mathématiques avec de nouvelles tables très exactes du soleil et de la lune* (Lausanne: A. Chapuis, 1754).

10. Utenhove points out (p. 178) that he had already remarked in notes to the previous Letter that such a project has no more meaning due to the work of Laplace and Lagrange on the mutual perturbation of planets.

11. The measure of imprecision of Ptolemy's catalogue defines a margin of error which in turn can only be coped with if a displacement of $\frac{1}{4}$ degree is assumed.

12. The correct number, as Utenhove remarks (p. 180), is 2182.

13. Utenhove refers (p. 181) to very recent data published in the *Connoissance des Temps* (1798, p. 221), according to which some less noticeable stars evidenced the largest proper motion.

14. As Utenhove reported (pp. 181–82), F. T. Schubert had done on the basis of more reliable data the same type of calculation concerning the displacement of stars since Ptolemy in his *Theoretische Astronomie* (vol. II, pp. 52–53; see note 180 to the Introduction), but what caught Utenhove's eyes was a minor error in calculation and not Schubert's failure to refer to Lambert, although the context was quite Lambertian.

15. See note XII-11.

16. What Lambert suggests here is that the same law cannot be extended beyond fourth-magnitude stars. See note XII-10.

17. In his note (p. 184) Utenhove refers to the conflicting specifications by Kant and by Herschel of the center of the Milky Way, but he does not point out that for Lambert the center of the local system of stars is not at all the center of the Milky Way.

18. Darquier refers (p. 185) to Messier's catalogue of nebulae published in 1783 as the fulfillment of that project, to which Utenhove hastens to add a reference to Herschel's vastly greater catalogues of nebulae.

19. Indeed, something very important with respect to Lambert's own cosmology as will be clear in Letters XVII and XVIII.

LETTER XIV

1. *Ovid Metamorphoses*, Latin original with an English translation by F. J. Miller (Loeb Classical Library; London: W. Heinemann, 1928), vol. I, pp. 8–9 (Lib. i, 85–88).

2. A paraphrasis of the opening phrase of Psalm 18.

3. For this very reason the subsequent analysis by Lambert of the vision of point objects has its inevitable shortcomings and ambiguities. To discuss them in detail in the context of eighteenth-century optics is not necessary for an appraisal of his cosmology.

4. The third division of a degree by sixty, or one sixtieth of a second.

5. Pieter van Musschenbroek (1692–1761) was professor of physics at Leyden. See his *Introductio ad philosophiam naturalem* (Leyden: apud Sam. et Joh. Luchtmans,

1762) vol. II, p. 732, with references there to previously printed accounts of this "experiment."

6. Darquier notes (p. 190) that according to Bouguer the corresponding figure is 300,000th. Darquier most likely refers to the second enlarged edition of Bouguer's work published posthumously in 1760 as *Traité d'optique sur la gradation de la lumière* (Paris: H. L. Guerin & L. F. Delatour); see p. 87. The first edition (see note 24 to the Introduction) had the same figure on p. 33.

7. Galileo presented his count in his *Sidereus nuncius* in 1610. See its English translation, *The Starry Messenger,* in Stillman Drake, *Discoveries and Opinions of Galileo* (Garden City, N.Y.: Doubleday, 1957), p. 48. Utenhove refers (p. 191) to several other counts of stars in Orion made shortly after Galileo, and expresses his surprise that in spite of that search of Orion, its nebula was not discovered until 1659. To Utenhove this suggested that the intrinsic visibility of that nebula was variable, a point which Lambert emphasizes in Letter XVIII.

8. Utenhove remarks (p. 191) that since the area explored by Galileo in Orion was much smaller, Lambert's calculation is much too modest and even "a bit arbitrary." because of this the estimated number of stars should be higher even when only telescopes of mediocre magnification are taken into account. With the best modern telescopes, Utenhove continues, the number in question should rise into the millions and he refers to Schröter, who spoke of 12 million stars within the reach of his 27-foot telescope (*Aphroditographische Fragmente,* p. 40), and to Lalande, who in his *Astronomie* specified as 75 million the number of stars that could be seen with Herschel's telescopes. See 3d revised edition, Paris: chez la Veuve Desaint, 1792, vol. I, p. 189.

9. For the previous occurrence of this series, see Letter XII and notes 10 and 11 there. Utenhove calls (p. 192) Halley the "inventor of the series" (for details, see my *The Paradox of Olbers' Paradox,* pp. 81–82), in one of the very rare references in this connection to Halley up to that time. Utenhove then remarks on the ambiguities involved in the determination of magnitudes and proposes the arithmetic series 1, 4, 7, 10, 13, which generates the series 1, 4, 12, 26, 51, the terms of which, when multiplied by 18, turn into the sequence, 18, 72, 216, 468, 918, in "marvelous agreement" wih the number of stars from the first to the fifth magnitude.

10. Utenhove points out (p. 192) that the number 1,650,120 would accommodate only stars up to the 74th magnitude whose exact sum is 1,653,900, and that Lambert's error was already being repeated, but the reference given by him is to an entirely insignificant textbook on practical astronomy.

11. The presence of an "inner diameter" in Lambert's notion of the Milky Way is invariably overlooked in the literature. The disk to which he had already compared the Milky Way has, therefore, a relatively large hole in its center and might be compared to a flat ring. Such a model, never considered by Kant, is one of the two models endorsed by Wright. See my *The Milky Way,* p. 190.

12. Utenhove recalls p. 193) that Herschel felt confident in his two great memoirs on the construction of the heavens (1784 and 1785) that he had fathomed the depths of the Milky Way. Subsequent observations of Herschel made him, however, conclude that the Milky Way was "unfathomable."

13. Utenhove refers (p. 194) to his previous note (XII-25) and recalls that Newton had already warned (*Principia,* Bk. III, Prop. XIV) that the fixity of aphelia and nodes could be modified by the mutual perturbation of planets.

14. Aries is the first in the sequence of the signs of the zodiac.

15. In his *Tabulae Rudolphinae* (1627).

16. As Utenhove notes (p. 196), it had become clear on the basis of more improved Tables that the motion of nodes with respect to the zodiac is retrograde for all planets. Consequently, Utenhove remarks, there is no need to look with Lambert outside the solar system for the explanation of some of these motions.

17. In a note which clearly undercuts Lambert's speculation, Utenhove notes (p. 196) that if Herschel's claim of the motion of the sun toward λ Herculis has any merit, then the inclination of the sun's orbital plane to the ecliptic cannot be less than the latitude of that star, or 49 degrees.

18. Another case of Lambert's hasty phrasing. He should have written, in clear analogy with the immediately preceding sentence: "rather than in opposition" instead of "and opposition."

19. Utenhove notes (p. 197) that Herschel had found in a direction opposite to Orion the greatest number of stars lying behind one another.

LETTER XV

1. According to Utenhove (p. 200) "Ferdinand de Magellan, famous Spanish navigator of Portuguese birth, the first European to sight the Pacific Ocean, set out on a tour around the globe, but perished as a victim of savage islanders he visited in the course of the voyage, and the tour was completed only by one vessel of his squadron which returned on September 7, 1522, to Europe from where it had departed on August 10, 1519."

2. The convenience of an invariable plane made many a prominent astronomer look for it. Darquier himself refers (p. 201) to Euler and D. Cassini. Utenhove adds that Laplace showed much concern for such a plane in the first volume of his *Traité de mécanique céleste*. This first volume was published in An VII [1798] (Paris: de l'Imprimerie de Crapelet). On the invariable plane, see Liv. i, chap. 7, § 62, pp. 317–21.

3. See note 18 to the Preface.

4. The paper was not published until 13 years after Mayer's death. See note 18 to the Preface.

5. By synthetical and analytical Lambert essentially means inductive and deductive, respectively.

6. This is the law of inertial motion, or Newton's first law.

7. The notion of gravity as pressure in the ether was originally endorsed by the Cartesians. See note III-7. Lambert's reference to pressure in the ether is expressive of the desire of all classical physicists to find a mechanical explanation of gravitation, a project about which Maxwell wrote a hundred years later that men of science would gladly devote the "whole remainder of their lives" to it, if a promising theory were available. See *The Scientific Papers of James Clerk Maxwell,* edited by W. D. Niven (Cambridge: University Press, 1890), vol. I, p. 156.

8. Such a rendering of the original which is demanded by Lambert's reasoning assumes that "von den übrigen Planeten" is an erroneous printing of the intended "und der übrigen Planeten."

9. This suggestion that the writer of Letters of odd numbers is familiar with the workings of the solar system but not with that of the system of stars is difficult to reconcile with the contents of the Letters in question. Whatever tediousness there is in the *Cosmologische Briefe* is in part due to Lambert's corresponding more with himself than with someone who really needs to be instructed. The literary success of Fontenelle's *Entretiens* was in large part due to the fact that the respective roles of instructor and learner were always carefully kept.

10. See note III-8.

11. Here (p. 206) both Darquier and Utenhove added a note. The former's remark, based on data gathered up to 1785, stated that the respective ratio was 51 to 16, whereas the latter, writing fifteen years later, gave the ratio as 58 to 20. Both expressed their satisfaction over the agreement of data with Lambert's theory.

12. In the subsequent Table these units of ten degrees refer to both hemispheres.

13. The + sign after three entries in the column of the computed number of comets indicate the excess due to the fact that the division of 21 comets into 9 units would yield $2\frac{1}{3}$ comets for each. This slight excess over 2 is taken into account by Darquier who gave a Table rather different in appearance though not in substance from the one in the original. In his note (p. 207) Utenhove tabulated 94 comets by adding 5 to the list of 89 given by Zach (see Introduction and note 159 there) and found a very good agreement with Lambert's theory.

14. The thrust of this conclusion, which is left out in Darquier's translation, is that after all there may be an even distribution of the 21 comets along the 12 signs of the zodiac. Utenhove, who did not report Darquier's omission, drew up a distribution of 94 comets and found that the statistics brought out curious discrepancies which he did not wish to dwell upon (p. 208).

15. The + and − signs in the column of computed numbers indicate the discrepancy of observed positions from the computed ones. Darquier once more gave a "corrected" version of Lambert's Table. Utenhove's tabulation of 94 comets gave with Lambert's Table an agreement "as good as one can desire" (p. 208).

16. According to Utenhove (p. 209), the count of 94 comets showed a perfect balance in this respect. 46 were in the northern, 48 in the southern signs.

17. According to Utenhove, of 94 comets 54 passed through their perihelia in winter months and 40 in summer months (p. 209).

18. Since of the 94 comets 49 were direct and 45 retrograde, Utenhove could write with obvious satisfaction: "The equality obtains even here in a marvelous manner" (p. 210).

19. The comets of 1337 and 1472.

20. Indeed, only one comet in that Table has a perihelion slightly beyond the earth's orbit.

21. 40 is the square of the ratio of the distances of Saturn and Mars from the sun. The number of comets increases, according to Lambert, as the square of distance from the sun. See Letter IV.

22. A good example of Lambert's imprecise phraseology. What he actually means, if his immediately following sentence is to fit the context, is that no cometary orbit has an angle of inclination which would be less than the maximum angle of inclination among planetary orbits.

23. At the end of this Letter Darquier submits a Table of his analysis of the angle of inclination of 69 cometary orbits, and after reporting a good agreement with respect

to the five other parameters between Lambert's theory and data three times more numerous than in Halley's Table, he concludes that Lambert's "principle has not lost any of its probability by a much more numerous combination [of data]" (p. 214).

LETTER XVI

1. See note 18 to the Preface.

2. *Ibid.*

3. See note 4 to the Preface and note IV-8.

4. As contrasted to "being in conjunction," that is, on the same side with respect to the center.

5. In Plato's *Timaeus* the Great Year's length is determined by the return of all planets to the same relative position. After the discovery of the precession of the equinoxes, the Great Year was usually equated with its period. For further details and for the whole history of the concept of Great Year in cosmological speculations, see my *Science and Creation: From Eternal Cycles to an Oscillating Universe* (Edinburgh: Scottish Academic Press; New York: Science History Publications, 1974).

6. The purpose is to populate the universe as completely as possible.

7. Lambert's readiness to assume densities, or specific weights, far exceeding that of gold, is worth noting in view of a possible comparison of his "dark bodies" with neutron stars and with the "black holes" of modern relativistic cosmology.

8. A plausible assumption if indeed comets were comparable in mass to planets. It is well to recall that the mass of Halley's comet is about a thousand millionth of the earth's mass, and other comets were found with masses many times smaller. Clearly, millions of comets would still be far from sufficient to counterbalance the sun's mass.

9. The concluding part of this phrase, characteristically omitted by Darquier because of its obscurity in the original, should most likely read: "and the position of the earth with respect to their orbits is more stable," that is, less subject to perturbations.

10. Because of this, comets will always remain broadly distributed in space. The incommensurability of planetary orbits was already emphasized by Nicolas Oresme in the latter half of the fourteenth century. See my *Science and Creation,* p. 237.

11. The + and − signs in the columns of computed numbers indicate a slight deviation from the integer. Typically, these signs are omitted in Darquier's translation.

12. Utenhove supplements (p. 227) Lambert's Table with one based on 94 comets and the results are the same: "The differences here, too, are all of the same sign and much too great for the assumption that planes of any direction are equally possible."

LETTER XVII

1. Darquier seems to have been taken aback by that enormous time-span, and perhaps to spare Lambert's French readers a similar impression, he rendered the original as follows: "Perhaps the century to which this information is reserved and which I have removed very far, is not too distant from ours" (p. 231).

2. Utenhove refers (p. 232) to Kant's advocacy of such an absolute center in the universe without pointing out that whereas such a center had a logical place in Lambert's finite universe, it was wholly illogical on Kant's part to say that whereas in infinite space any point could be taken for center, the evolution of the infinite physical

universe had to start from one specific center and that singling out such a physical center was not at all contradictory. See *Allgemeine Naturgeschichte*, p. 110.

3. The reference is to the view according to which the satellites of Jupiter became attached to it through capture. This view had never been argued systematically by anyone. It was rather an implicit corollary of the role assigned to near-collisions by Whiston in the formation of the planetary system. See notes I-5 and V-2.

4. The simile is worth noting because later Lambert will use comparisons equally political in character which also indicate his marked conservatism.

5. This is the first appearance of a term which will be used by Lambert more frequently and emphatically.

6. Satellites or moons.

7. Utenhove refers (p. 235) to his reasons given in connection with some enormous stars mentioned in Letter XII (note 7 there). What Utenhove seems to have in mind is that because of the very strong gravitational field of those dark bodies, no light would be reflected from them.

8. Utenhove tabulates 94 comets in two groups, southern and northern, and concludes that "comets in general are more numerous and more crowded toward the zodiac than elsewhere" (p. 238).

9. The greater number of observations made it evident already by 1800, as shown by Utenhove in the preceding note, that Lambert's claim was not to be borne out by the accumulation of further data.

LETTER XVIII

1. Utenhove remarks (p. 242) that with the exception of Herschel no leading astronomer was willing to see that nebula as another Milky Way composed of great many stars. But, Utenhove adds, none of those astronomers was inclined to take with Lambert that nebula for a dark central body. More correctly, no leading astronomer had cared up to that time to comment on Lambert's dark bodies. The greater part of this long note of Utenhove is taken up by his account of Kant's specification of Sirius as the central star of the Milky Way.

2. On Derham, see note IX-1. Contrary to Lambert's suggestion, there is not the slightest indication in Derham's paper that he had viewed the light seen through those openings as "reflected" light.

3. See note 41 to the Introduction. A telling evidence of the liberties taken by Darquier occurs at this point (p. 243) where in the text Huygens is mentioned and his discovery of the great nebula in Orion. According to Darquier's note the changes in its contours were very noticeable in 1782.

4. Most likely Lambert refers to the two engravings in de Mairan's treatise on the aurora borealis (see note 41 to the Introduction; 2d ed., Planche XV, Fig. XXVI & XXVII).

5. Utenhove once more refers to the optical paradox of such a large body, intrinsically luminous or not, as pointed out by Laplace. See note 178 to the Introduction.

6. Utenhove reports (p. 246) a recent communication by E. Prosperin to Zach's *Monatliche Correspondenz*, 1 (1800), pp. 117–18, where Prosperin stated that according to his studies the satellites of Jupiter would go into a hyperbolic orbit around the sun had Jupiter suddenly ceased to attract them. To this Prosperin added

that there was no serious possibility for a comet to become a satellite of Jupiter. In all this, Prosperin did not refer to Lambert, a point not mentioned by Utenhove.

7. Such is certainly a most sound directive for an astronomer or cosmologist. Any major advance in penetrating deeper into the universe has invariably unveiled unsuspected features of it (see on this Chapter V, "The Frontiers of the Cosmos," in my *The Relevance of Physics* [Chicago: University of Chicago Press, 1966], 188–235). Yet, too much boldness in speculating about still unexplored realms of the cosmos can be very misleading.

8. Bodies of the first and second rank are the satellites and the planets, respectively.

9. Whereas Lambert speaks of creation, Darquier has Creator (p. 247).

10. Utenhove points out (p. 248) that it would have been more exact on Lambert's part to speak of the moon's path as an epicycloid since a cycloid is traced out by any point of a circle rolling along a straight line. Utenhove refers furthermore to Laplace's *Exposition* (Liv. v, chap. 6) where this difference is emphasized not, of course, with any reference to Lambert, a point once more overlooked by Utenhove.

11. In the strictly hierarchical system of Lambert the concept of democracy has, of course, no place even as an analogy. Furthermore, since this hierarchical system is a perennial cosmic establishment, democracy in it could only mean an evolution tantamount to revolution.

12. Because of our inability to ascertain the rank of the absolutely central body.

13. Lambert should have written "Milky Ways and so forth," for a cycloid of second degree is described by systems which are only two steps removed from the absolutely central body. They are at such an inconceivable distance from us that thousands of centuries, or nearly a million years, might be needed to learn anything positive about them.

14. The conjunction of Jupiter and Saturn, which occurs three times during two full revolutions of Saturn.

15. The finding of such reasons can only be done in an evolutionary perspective, an anathema for Lambert.

16. Taking the zodiacal light for the outermost part of the atmosphere of the sun was a procedure common with most astronomers at that time. Utenhove remarks that Laplace showed in his *Exposition* (Liv. iv, chap. 9) and in his *Traité de mécanique céleste* (Liv. iii, chap. 7) that the solar atmosphere can at most extend to the orbit of Venus. To this Utenhove adds two explanations of the zodiacal light proposed by Kant in his *Allgemeine Naturgeschichte,* admitting, however, that "the true cause of this phenomenon is still unknown to us and we have but more or less probable conjectures about the material which constitutes the zodiacal light" (p. 254).

17. See notes 59 and 60 to the Introduction and note IX-8.

18. A typically unclear and repetitious phrasing.

19. The reason as given shortly below by Lambert is that the zodiacal light does not stretch all over the sky.

20. It is known today that the zodiacal light resembles in its extent a flat spheroid which extends beyond the earth's orbit in every direction and is concentric to the sun. An evidence of this is the so-called Gegenschein, or a very faint light opposite to the zodiacal light. The zodiacal light is due to the reflection of light from particles accrued from the fragmentation of asteroids and of comets drawn toward the sun.

1. Here (p. 257) Darquier inserted a quotation, "quantum mutatus ab illo," from Vergil's *Aeneid* (Lib. ii, 274), as if it stood in the German original. Moreover, Darquier's rendering of it has no reference to languages and he presents the resting of the centralmost body as an "absolute" one. With so many liberties taken with the original within one brief paragraph the result is not a translation but a paraphrasis.

2. The allusion is to Tycho Brahe's introducing two centers into the solar system by letting Mercury and Venus revolve around the sun which is in turn orbiting around the earth.

3. A clear indication that in speaking of the centralmost body as being of the 1000th rank Lambert spoke figuratively.

4. The exact figure is 105 as Utenhove notes (p. 260).

5. Utenhove refers (p. 262) to investigations by Euler and others who showed that the uniformity of the earth's rotation was the result of the mutual compensation of various inequalities affecting it. Utenhove also refers to some exceedingly minor inequalities which Laplace found in the earth's rotation. The earth's rotation has in very recent times been superseded by atomic clocks as the basis of reference in measuring time.

6. See note XVI-5.

7. See note XVIII-2.

8. Actually there are stars of such size but they are not at all solid!

9. Darquier notes: "This is no longer true since the discovery by Herschel [of Uranus]" (p. 266).

10. This angle, Darquier notes, can be magnified to about 2 minutes with Herschel's telescopes (p. 267).

11. A fairly good prognosis. According to present-day classification, the "long-period" class of comets begins with a period of 200 years. Only a few comets are known to have periods of a few thousand years. Estimates of much higher periods are beset with great uncertainties.

12. Utenhove explains (p. 267) this law with reference to the inverse ratio of the attractive force to the square of distance.

13. See note IX-1.

14. A recurring series, as Utenhove notes (p. 268), is a sequence of numbers in which each term is the sum of some of the immediately preceding terms multiplied by a constant. First considered by De Moivre, they were extensively studied by Euler in his *Introductio in analysin infinitorum* (1748). Utenhove also notes that there is an infinite variety of such series, one of them being the well-known geometrical progression.

15. See de Mairan's treatise (note 41 to the Introduction; 2d ed., pp. 31–40).

16. Utenhove identifies one such astronomer as "the Jesuit Father Noel" (p. 270). François Noel (c. 1649—c. 1725), Jesuit missionary and astronomer, was the author of *Observationes mathematicae et physicae in India et China factae ab anno 1684 usque ad annum 1708 una cum mappa stellarum australium* (Prague, 1710).

LETTER XX

1. A reference to the growing influence of empiricism advocated especially by Hume, to the reaction to Cartesian a priorism prompted by the success of Newtonian physics, and to the general decline of Wolff's influence.

2. Utenhove notes (p. 276) that in spite of the inherent "invisibility" of such a large body, its existence cannot be accepted without some sort of experimental verification; that its existence is not absolutely necessary for securing the orderliness of a system of stars; that Newton had ascribed the apparent immobility of stars to their homogeneous distribution in space (*Principia,* Book III, Prop XIV); that Herschel did not find the existence of such a body necessary; and that it had recently been pointed out by F. Soldner (*Astronomisches Jahrbuch für das Jahr 1803* [Berlin, 1800], p. 190) that Herschel conducted his observations in the belief that the sun was not far from the center of the Milky Way. Such a central position of our sun, Utenhove adds, should make the motion of stars more conspicuous than suggested by observations. Utenhove does not, however, note that neither Soldner, nor Herschel spoke of that problem with a reference to Lambert.

3. Utenhove refers (p. 279) to the great probability of the formation of a very large body in the center of stellar systems, as suggested by Kant, but once more Utenhove points out both the absence of rigorous mathematical necessity for this to happen and he once more recalls the remarks of Soldner mentioned in the preceding note.

4. Lambert's ability to distinguish clearly between Newton's ideas on gravitational attraction and those of his followers is certainly remarkable when taken in the context of the times. Utenhove, however, represents (p. 280) a facile formalism incompatible with the breadth and width of Newton's method when he claims that "in order to be a disciple of that immortal man, it is enough to admit that attraction acts according to constant laws which he had established, without pretending how it acts, that is, for example, through the mediation of a certain matter, or not: it has been for a long time that geometers abandoned to metaphysicians the concern for such investigations" (p. 280). The fact is, however, that Newton's method had always remained deeply rooted in epistemological tenets that implied genuine metaphysics. See on this the sixth of my Gifford Lectures (University of Edinburgh, 1975 and 1976) to be published soon under the title, *The Road of Science and the Ways to God.*

5. According to a note by Utenhove (p. 282), this is the opinion of Lalande, who follows Maupertuis "and most of the English metaphysicians." Indeed, the latter turned gravity into an intrinsic, innate quality of matter, largely under the influence of Samuel Clarke, Newton's spokesman in his dispute with Leibniz.

241

INDEX OF NAMES